# The Interface Envelope

# The Interface Envelope

## Gaming, Technology, Power

**JAMES ASH**

Bloomsbury Academic
An imprint of Bloomsbury Publishing Inc

B L O O M S B U R Y
NEW YORK · LONDON · OXFORD · NEW DELHI · SYDNEY

**Bloomsbury Academic**
An imprint of Bloomsbury Publishing Inc

| 1385 Broadway | 50 Bedford Square |
| New York | London |
| NY 10018 | WC1B 3DP |
| USA | UK |

www.bloomsbury.com

**BLOOMSBURY and the Diana logo are trademarks of Bloomsbury Publishing Plc**

First published 2015
Paperback edition first published 2016

© James Ash, 2015

All rights reserved. No part of this publication may be reproduced or transmitted in any form or by any means, electronic or mechanical, including photocopying, recording, or any information storage or retrieval system, without prior permission in writing from the publishers.

No responsibility for loss caused to any individual or organization acting on or refraining from action as a result of the material in this publication can be accepted by Bloomsbury or the author.

**Library of Congress Cataloging-in-Publication Data**
Ash, James, 1983-The interface envelope : gaming and the logics of affective design / James Ash.pages cmSummary: "Develops the new theory of 'interface envelope': how interfaces produce envelopes of space and time that serve to focus users' perception on the present moment"-- Provided by publisher.
Includes bibliographical references and index.
ISBN 978-1-62356-459-9 (hardback) 1. Video games--Psychological aspects. 2. Video games--Design 3. User interfaces (Computer systems) I. Title.
GV1469.39.P79A75 2015794.8'1536--dc23
2014034651

ISBN: HB: 978-1-6235-6459-9
PB: 978-1-5013-2000-2
ePub: 978-1-6235-6557-2
ePDF: 978-1-6235-6975-4

Typeset by Integra Software Services Pvt. Ltd

*For Calum*

# Contents

Acknowledgements  ix

**1** Introduction  1
  Attention, economy, power  4
  Post-phenomenology and new materialism  7
  Media, software and game studies  9
  Chapter outlines  11

**2** Interface  15
  Interface theory  16
  Interfaces as objects  25
  Interfaces as environments  29
  Interface, object, transduction  31

**3** Resolution  33
  Resolution  34
  Neuropower  37
  High and low resolution  40
  Phasing between resolutions  47
  Resolution, habit, power  54

**4** Technicity  57
  Technicity  59
  Psychopower  64
  Homogenization  67
  Irreversibility  73
  Technicity, time, power  78

**5 Envelopes** 81
Homeomorphic modulation 84
Envelope power 90
Shifting logics of the envelope in games design 95
The contingency of envelopes 103

**6 Ecotechnics** 105
The ecotechnics of care 109
Ecotechnics of care: Two sites of transduction 111
From suspended to immanent ecotechnical systems of care 115
The temporal deferral of negative affect 117

**7 Envelope Life** 119
Gamification 120
Non-gaming interface envelopes 122
Questioning envelope life 129
Pharmacology 137

**8 Conclusions** 139
Games/digital interfaces 140
Objects/new materialism/envelopes 143
Power/economy/capitalism 146

Bibliography 149
Index 163

# Acknowledgements

This book is the result of several years of thinking about interfaces, objects, affect, space, time, materiality and gaming. In many ways the book is interdisciplinary, sitting at the crossroads between cultural geography, media and cultural studies, game studies and media and social theory. Writing from an interdisciplinary perspective, the explicit aim of this book is to create a series of concepts that theorize interfaces by linking aspects of these diverse areas of literature together. Drawing upon geographical ideas of space and time, media theorizations of interfaces and cultural understandings of materiality, I hope the book creates productive connections that can speak to and benefit all of the above audiences.

The writing of the book has benefitted from the support of many friends and colleagues, both directly and indirectly. I would like to thank JD Dewsbury, Phil Crang and Gillian Rose, who offered advice and institutional support at key moments in the past. My sincere thanks go to Pepe Romanillos, Richard Carter White, Charlie Rolfe, Sam Kinsley and Paul Simpson. As colleagues in the School of Geographical Sciences at Bristol University between 2005 and 2009, they set a very high benchmark and encouraged me to improve and develop my own work. I also want to thank Pepe for giving me *Technics and Time 1* as a birthday present in 2005. I had never encountered Stiegler before and reading this text introduced me to phenomenology, which had a transformative effect on my thinking. More broadly, I would like to thank Paul Harrison, Ben Anderson, Matt Wilson and John Wylie, who have offered sage advice relating to various aspects of academic life. Their openness and generosity has helped on a number of occasions and I hope I can dispense equally good advice to others in the future.

The book has benefitted from audiences who have heard and responded to versions of various chapters at seminars between 2012 and 2014. Versions of Chapter 6 have been given at Newcastle University and Royal Holloway Geography departments, and aspects of Chapter 3 were given at the Conditions of Mediation ICA pre-conference at Birkbeck University in London. I would also like to thank Peter Stone at Newcastle University, who approved a semester of study leave that enabled me to finish and submit the manuscript. Finally, thanks to Pepe Romanillos, Paul Simpson and Lesley Gallacher for reading versions of the manuscript and offering advice. All errors remain my own.

# 1

# Introduction

*As the window of a Web browser replaced cinema and television screen, the art gallery wall, library and book, all at once, the new situation manifested itself: All culture, past and present, came to be filtered through a computer, with its particular human-computer interface.*

MANOVICH (2001: 64)

*[the interface is]… the place where flesh meets metal or, in the case of systems theory, the interface is the place where information moves from one entity to another, from one node to another within the system.*

GALLOWAY (2009: 936)

As Lev Manovich argues, human life in the western world is increasingly mediated by digital, computational interfaces. Computers, laptops, tablet PCs, mobile phones, videogames and many other devices operate as the medium through which a variety of activities are undertaken. In a general sense, as Alexander Galloway suggests, the term 'interface' refers to a surface or point of contact between two entities. These can be between a human being and a technical object (in the case of a tool) or between two technical objects (in the case of software). As such, an interface can be a door handle, a steering wheel or a baseball bat as much as graphical user interface on a personal computer or a key on a keyboard. Further reflecting on the concept of interface in relation to software, Galloway (2009: 936) goes on to argue:

> the interface becomes the point of transition between different mediatic layers within any nested system. The interface is an 'agitation' or generative

friction between different formats. In computer science, this happens very literally; an 'interface' is the name given to the way in which one glob of code can interact with another. Since any given format finds its identity merely in the fact that it is a container for another format, the concept of interface and medium quickly collapse into one and the same thing.

Any serious consideration of what an interface is opens up a broader question about the relationship between concepts of the medium, object and technology. Upon closer examination an interface appears to be a medium itself, a piece of technological equipment as well as an object. Understandings of the interface (digital or otherwise) as a surface or space of contact are also loaded with specific assumptions about how objects, whether natural or technical, encounter and relate to one another in general. Indeed, questions about what an object is, what a technical object is and how these objects relate to one another are a central concern of philosophy itself. Famously, Martin Heidegger (1982: 4–5) argued that technology has been understood from an 'instrumental perspective':

> we ask the question concerning technology when we ask what it is. Everyone knows the two statements that answer our question. One says: Technology is a means to an end. The other says: Technology is a human activity. The two definitions of technology belong together. For to posit ends and procure and utilize the means to them is a human activity. The manufacture and utilization of equipment, tools and machines, the manufactured and used things themselves, and the needs and ends they serve, all belong to what technology is. The whole complex of these contrivances is technology.... Who would ever deny that it is correct.

Drawing upon Heidegger's definition, interfaces can be understood as mediums, objects and technologies. They exist in order to fulfil operations and, thus, are a means to an end; they are created through the human shaping of matter and also operate as a tool for human action. Despite his own criticism of the definition, Heidegger's (1982: 20) instrumental account remains prevalent. Implicit in this narrative is an assumption that objects are ultimately substances with specific properties and that what differentiates technology from naturally occurring objects is the extent to which these properties have been manipulated by humans. For example, a stone is a natural object that has been produced by non-human processes such as erosion or weathering, while a flint is a technical one, because the flint has been broken by humans against other stones to create a sharpened point.

This book demonstrates that, as technical objects, interfaces can be understood in ways that are not reducible to this instrumentality. By rethinking categories surrounding 'matter', 'objects' and 'technology', the chapters that follow develop a series of concepts in order to understand human practices with interfaces as potentially productive of what I term 'spatio-temporal *envelopes*'. For now, envelopes can be defined as localized foldings of space-time that work to shape human capacities to sense space and time for the explicit purpose of creating economic value for the designers and creators of these interfaces. Through examining a number of contemporary videogames, I argue that creating economic value in this way can be understood as a new form of power, which I term *envelope power*.

Videogames are a useful site to theorize these interface envelopes for a number of reasons. First, videogame interface design is subject to rapid and continual change and so offers a window into the technical historicity of the interface. The past twenty years have seen huge innovations in how information is conveyed to players in videogames. Games consoles have moved from simple, single-button joysticks to a situation in which a large number of players now interact with games using remotes and cameras that can register complex forms of analogue bodily movement. Videogames, therefore, offer a wealth of examples of different Graphical User Interfaces (GUIs) all of which have different logics and mechanics (Jørgensenn 2013). This allows us to study different videogames and analyse the different envelopes they attempt to generate.

Second, videogame interfaces are at the forefront of interface technology, both in terms of the devices used to control videogames and in terms of the software engines that run the games themselves. The Nintendo Wii gestural remote, the Playstation Move gestural remote and camera and the Xbox Kinect camera all radically reframe how players interface with games consoles. Technologies of gestural interface existed before these systems, but videogames have led the way in integrating these systems into mass-produced consumer devices (Jones and Thiruvathukal 2012).

Third, the logic of videogame interfaces is spreading to other areas and forms of life outside of games. Videogames can be considered as a barometer for broader developments in interface design. For example, in a promotional 'vision video' (Kinsley 2010) for their Kinect camera sensor, Microsoft (2011) points to a range of non-gaming applications for the device. The video depicts a range of scenarios including a surgeon using the Kinect to manipulate visual patient files without having to re-sterilize themselves and a child using game-like programmes as part of physiotherapy. While these are 'visionary' rather than actual applications of the technology, videogames offer a window through which to understand the multiple forms, logics and possible futures of the digital interface more generally.

We can use the concept of the envelope to identify shifts that are occurring in videogame design and understand the effects of these envelopes on players' spatio-temporal perception. There is a trend in current big-budget videogame design and publishing (often referred to as AAA or triple A games) towards attempting to encourage players to concentrate on a modulating present moment, in an increasingly narrow spatio-temporal envelope of perception. The ways in which videogame interfaces are designed are central to the production of these narrow envelopes. A theory of interface envelopes can help us to critically reflect upon the forms of 'presentness' that videogames enable, how this presentness shapes the reflective and critical thinking of players and the extent to which this presentness may be spreading to other types of non-videogame interface.

## Attention, economy, power

The concepts of envelopes and envelope power can be understood in the context of emerging claims around how interfaces operate as the site of new economies and forms of power. Writers such as Bernard Stiegler (2010a), Katherine Hayles (2007) and Douglas Rushkoff (2013) argue that the content presented through digital interfaces, such as television, videogames and social media, is creating a form of distracted present in which human capacities for imagination, long-term thinking and careful reflection are being broken down. Rather than developing 'deep' modes of attention, based around temporally elongated activities such as reading, these industries create a 'hyper' attention, where increasing levels of stimulation are required to keep viewers interested in a single subject or topic (Hayles 2007). In Rushkoff's (2013: 3–4) estimation, the digital world is creating a 'multitasking brain, actually incapable of storage or sustained argument', leading to what both he and Stiegler (2009a) term a 'temporal disorientation'. Instead of gaining pleasure from focusing on one activity, 'we hop from choice to choice with no present at all. Our availability to experience flow or to seize the propitious moment is minimised as our choices per second are multiplied by a dance partner who doesn't see or feel us' (Rushkoff 2013: 115). In turn 'we lose the ability to imagine opportunities emerging and excitement arising from pursuing whatever we are currently doing, as we compulsively anticipate the next decision point' (Rushkoff 2013: 116). This breakdown of narrative, which Turner (1998) argues is central to how we interpret, understand and make sense of our lives and the world, has a profound effect; it creates what Rushkoff terms a 'perpetual present tense' or 'perpetual now'.

This form of presentness should not be confused with a positive state of what Csikszentmihalyi (2009) has called flow, or as some form of digital mindfulness. In Rushkoff's (2013: 4) words:

> we are not approaching some Zen state of an infinite moment, completely at one with our surroundings, connected to others, and aware of ourselves on any fundamental level. Rather we tend to exist in a distracted present, where forces on the periphery are magnified and those immediately before us are ignored. Our ability to create a plan – much less follow through on it – is undermined by our need to improvise our way through any number of external impacts that stand to derail us at any moment. Instead of finding a stable foothold in the here and now, we end up reacting to the ever present assault of simultaneous impulses and commands.

Beyond individual negative effects, Stiegler suggests that, in attempting to control attention, these so-called 'cultural industries' also create problems for the construction of social relations. The issue with attention controlling apparatuses such as social media is 'that ... [they] ... destroy ... attention itself, along with the ability to concentrate on the object of attention, which is a social faculty; the construction of such objects is in fact the construction of society itself, as civil space founded on cultural knowledge, including social graces, expertise and critical thinking (i.e. contemplation)' (Stiegler 2010a: 13). To illustrate this, Stiegler (2013a: 82–83) gives the example of families who all sit separately in front of their own screens and so do not spend time learning to identify or relate to one another. Stiegler argues this is problematic because parents do not pass on somatic skills such as empathy or care to their children, which he considers to be the skills that are key to forming strong communities and societies.

In 'Taking Care of Youth and the Generations', Stiegler (2010a) terms this process of capturing and holding attention by market forces an 'attention economy'. Beller (2006: 4) suggests that attention is 'the newest source of value production under capitalism today'. For Beller (2006: 4), reality itself becomes organized around the production of a 'cinematic' form of attention: 'The cinematic organization of attention yields a situation in which attention, in all forms imaginable and yet to be imagined ... is that necessary cybernetic relation to the socius – the totality of the social'. Attention becomes both a finite, exchangeable commodity and a necessary relation that offers access to others in the world. As Goldhaber (1997: n.p.) puts it 'having attention is very, very desirable, in some ways infinitely so, since the larger the audience, the better. And, yet, attention is also difficult to achieve owing to its intrinsic scarcity. That combination makes it the potential driving force of a very intense economy'. Put simply, the question of attention is important because, as Stiegler argues, the capture of attention through a variety of 'psychotechniques' is central to the monetization of audiences through advertising, the development of brand loyalty and the blurring of boundaries between work and play in a variety of media (Stiegler et al. 2007, Stiegler 2009a: 38, Yee 2009, Paul 2010, Crogan and Kinsley 2012).

From this perspective, videogame interfaces can be understood as part of a 'global cultural industry' (Lash and Lury 2007, Kirkpatrick 2013) and an example of the latest state of what has also been termed 'cognitive capitalism' (Brophy 2011, Peters et al. 2011, Boutang 2012), the explicit aim of which is the manipulation of perception in order to create economic value. In the case of videogame interfaces, positive affect and attention have economic value that designers attempt to generate to keep players engaged with the game and willing to consume further iterations of that game in the future. However, it is not straightforwardly the case that these industries work to create a 'perpetual now' (Rushkoff 2013: 261) in order to create profit; the techniques utilized in videogame and interface design attempt to modulate players' capacities for recollection and anticipation within envelopes of differing length. In this way, interfaces can be understood as generating a *continuously modulating now*, in which the ability to reflect on the past and anticipate the future are opened and closed in ways that are specific to this or that envelope. I term this form of productive and active modulation *envelope power*. As I outline in Chapters 3, 4 and 5, envelope power is influenced by, but distinct from, what Stiegler (2010b) terms psychopower (the specific capacity of media technologies to capture and hold attention) and Neidich (2013) refers to as neuropower (the construction of synaptic and habitual relations in the body and brain in order to influence consumer decision-making). In contrast, envelope power is the capacity of interfaces to organize the relationship between memory and anticipation through the production of localized foldings of space-time in order to generate economic value.

Envelope power relies upon the active contingency and the productive skills consumers generate through practices of consumption. While videogame mechanics can be developed to encourage users to hold negative affects in their bodies for long periods with negative implications (see Chapter 6), at the same time, the envelope power videogames generate also sensitizes players' bodies to cultivate new capacities to sense difference between increasingly small units of space and time. The 'microeconomy of attention' (Stiegler 2010a: 94) that envelope power tries to tap into is not about mobilizing a pre-given attentional structure, or creating docile bodies

(Foucault 1977), but about actively opening up and creating new capacities for attention and affect that can be mined in order to realize new forms of embodied and habitual value.

Investigating and questioning Stiegler and Rushkoff's critique of digital presentness is important, not just because of the popularity of videogames, but because the logics that underlie these forms of game design are bleeding out into broader processes and technologies in everyday life. Interface envelopes are not just limited to videogames but can be used to understand other interfaces and how they might shape human capacities to sense space and time. Analysing concrete examples of these technologies in action (as I do in Chapter 7) allows us to explore the extent to which they encourage the emergence of new critical faculties of response to attentional economies, or whether these capacities simply draw players further into circuits of consumption linked to these interfaces, as writers such as Steigler and Rushkoff assume.

## Post-phenomenology and new materialism

To discuss envelopes and envelope power, the book develops a series of concepts that are informed and influenced by work in phenomenology (Heidegger 1962), speculative realism (Bryant et al. 2011, Harman 2013, Gratton 2014), object-orientated ontology (Bogost 2012) and new materialist theory (Bennett 2009, Roberts 2012, Lapworth 2013) to form what could be termed a 'post-phenomenology' (Ihde 2008, 2010, Ash and Simpson 2014). Although a very broad term, phenomenology is a school of thought that attempts to describe the world through the ways in which it appears to the human being without recourse to prior theories (scientific or otherwise) that would purport to explain such experience (Merleau-Ponty 2002: vii–x). Drawing upon and critiquing phenomenology, writers in speculative realism, object-orientated ontology and new materialism share an enthusiasm for rethinking basic questions around what matter and objects are. For example, Graham Harman (2002: 19) argues that 'inanimate objects are not just manipulable clods of matter, nor philosophical dead weight best left to positive science. Instead, they are more like undiscovered planets, stony or gaseous worlds which ontology is now obliged to colonise with a full array of probes or seismic instruments – most of them not yet invented'.

This desire for a renewed engagement and exploration of objects emerges from recent critiques of philosophy, which argue that traditional phenomenology (amongst many other philosophical schools) is trapped within what Quentin Meillassoux terms a 'correlationist' perspective.

Meillassoux (2010: 5) defines correlationism as a perspective that argues 'we only ever have access to the correlation between thinking and being, and never to either term considered apart from one another'. In other words, any knowledge or experience of the world is partial, situated and human. Harman argues that the result of this is that accounts of matter, or objects in the external world, have generally been reduced to how they appear to consciousness. As Harman (2005: 16) puts it, 'the object is stripped of all independent power and considered only insofar as it flares into human view'. Stripped of this power, objects become understood as forms of inert, passive matter that only become powerful when manipulated or used by humans.

Work in object-orientated ontology, speculative realism and new materialism offers a challenge to phenomenology to think about how sensory experience is shaped by the world in ways that aren't reducible to or necessarily experienced by consciousness. For contemporary studies of technology and gaming, new materialism further reinforces a desire to explore the 'inter-object' relations that take place in complex technical systems on their own terms. This matters because the specificity of these systems has powerful effects on those who engage with them. A post-phenomenology develops ways of thinking that can attend to these inter-object relations and how they shape human capacities outside of the phenomenal realms of the subject. What is interesting about new materialism, speculative realism and object-orientated ontology as emerging schools of thought is a desire to move beyond commentary on others' work to create new concepts and ways of thinking about the world. With this desire in mind, the chapters of the book are organized conceptually in order to provide a tool kit to begin to think about the power of interfaces outside of the ways humans engage with them.

Each chapter develops a concept introduced or informed by a theorist who in different ways is responding to the phenomenology of Martin Heidegger. Heidegger is a key figure in the phenomenology of technology (Borgmann 2009, Ihde 2010) and his work has also had a large influence on contemporary theorists thinking about the nature of objects (Harman 2002). As such, Heidegger's thinking and those that develop his work offer a good starting point for conceptualizing the interface. Some of the theorists I draw upon who are influenced by Heidegger include Bernard Stiegler, Graham Harman, Peter Sloterdijk, Catherine Malabou and Jean-Luc Nancy. All these theorists have a different, complex and specific relationship with Heidegger's thinking and ideas. The concepts that organize each chapter begin from these thinkers but then move into new and, I hope, fertile ground by drawing upon other figures who are not as directly influenced by Heidegger, such as Ian Bogost, Bruno Latour and Warren Neidich. With this in mind, the concepts developed in the book should not be considered as strictly Heideggerian or coherent to the internal logic of Heidegger's writing. In the same vein, the book does not attempt to produce a meta-theory or account that reconciles the differences between these thinkers.

## Media, software and game studies

Envelopes emerge from the relationship between the user's body and what I term an 'interface environment'. Interface environments are composed of technical objects, and as such, envelopes are generated through relations between non-human objects and human practices. While the concepts I develop across the book emphasize the autonomous existence of objects outside the way they appear to humans, this should not be confused with a 'technological determinist' perspective in which technical objects become either 'the agent of social change' (Murphie and Potts 2003: 11, Smith and Marx 1994) or determine what a user or audience thinks or feels when he or she experiences a piece of media. Making the claim that the objects that constitute interfaces have an autonomous power beyond the humans that design, manufacture and use them is not the same as arguing that these objects determine what people do with them. Mark Hansen (2006a: 299) posits that:

> there is simply no such thing as technical determinism, not because technics don't determine our situation, but because they don't (and cannot) do so from a position that is outside of culture; likewise, there is no such thing as cultural constructivism understood as a rigid, blanket privileging of ideology or cultural agency—not because culture doesn't construct ideology and experience, but because it doesn't (and cannot) do so without depending on technologies that are beyond the scope of its intentionality, of the very agency of cultural ideology.

In other words, an account of technological determinism assumes that humans and technology are separate in the first place, which would allow one to determine the other. This is not the case. As a growing body of writers, such as Kittler (2009) and Stiegler (1993, 1998), now recognize the human and the technical have co-evolved alongside one another and so cannot be thought apart. Work in software studies and new media is beginning to recognize the co-evolution of the human and technical and the importance of thinking through complex forms of technical systems on their own terms (Parikka 2010, Berry 2011, Chun 2011). For example, Kirschenbaum (2008) writes about magnetic force microscopy, a technique used to analyse and record images of inscription on computer hard disks. He points to the radical contingency and singularity of data as it exists in inscriptions on a magnetic surface of the hard disk: 'as a written trace, digital inscription is invisible to the naked eye but it is not instrumentally undetectable or physically immaterial. Saying so is not a theoretical proposition but a discernible fact, proven by the observable behavior of some 8.5 million terabytes of storage capacity bought to market in one recent year alone' (Kirschenbaum 2008: 74).

Emphasizing the role that technical objects play in shaping and informing practices does not mean users are reduced to puppets that uncritically accept the content presented to them. This is not a return to the kind of media effects approach (Sherry 2004, Slater 2007, Bryant and Oliver 2009) that has been so heavily critiqued by writers in cultural studies (Barker and Petley 2002, Moores 2012). As I argue in Chapter 5, envelopes are homeomorphic, which means that they require engagement from a user to exist at all. Instead of trying to determine what users do or think, interface envelopes productively draw upon the contingency of the indetermination of users' actions in order to create envelope power. Rather than try to create 'preferred readings' (Morley 1993) or 'decodings' (Hall 1993) of a piece of media, envelopes aim to create '"shared sensoria" – affective, perceptual and cognitive temporal and spatial blocks that allow for specific processes of individuation' (Terranova 2007: 142). As such, envelope power is not about oppressing or dumbing down a user's critical faculties, but about actively amplifying their embodied capacities within a set of possibilities that the interface attempts to prioritize or naturalize.

An account of envelopes and envelope power, then, fills a gap in work on software and game studies that has tended to play down the role of the body and how software involves, invokes and shapes particular corporeal capacities. It is a simple observation that players need their bodies to experience and control the spaces and places of videogames. However, as Behrenshausen (2007: 335–336) argues, until recently academics have 'exhibit[ed] a near-exclusive preoccupation with video games' relation to players' embodied sense of sight at the expense of exploring other powerfully carnal modes of player–game engagement'. As Kirkpatrick (2009) points out, bodily sensation in relation to games is under-discussed. In his words '[it is] ... curious that the details of ... what each game feels like in the hand, so to speak—is so rarely a matter for reflection' (Kirkpatrick 2009: 130).

Keogh (2014) argues that this ignorance of the body has been due to a focus on studying games at a purely formalist level, with writers such as Aarseth (1997), Wolf (2001) and Juul (2005) attempting to develop theoretical concepts and systems to classify and analyse what makes videogames unique as a medium. In response to this formalism, Apperley and Jayemane

(2012) have suggested that a material turn has taken place in game studies towards a consideration of the relationship between body and interfaces. This literature examines the software and hardware of particular videogame platforms (Montfort and Bogost 2009), their relation to expression (Bogost 2007, 2011) and also how games are experienced on an embodied level (Giddings 2007, Crogan and Kennedy 2009, Giddings 2014). But, while this materialist turn in gaming has begun to examine embodiment, these accounts tend to ignore the political implications of embodied play. For instance, writing in relation to *Call of Duty*, Crick (2011: 266) suggests: 'when I "enter" the virtual world of a FPS such as Call of Duty 4, my experience is not one of disembodied perception nor can my body be reducible to a mere set of eyeballs. For example, sometimes my heartbeat races or my body feels rushes of excitement and jolts during moments of intense combat with NPCs [Non Player Character's]'. In this case, Crick focuses on playing *Call of Duty* as a particular, individualized experience, and in doing so does not link the bodily capacities and senses invoked through the game back to broader infrastructures of power or politics.

When power or politics is invoked to discuss games, this has often taken place on a representational level to examine how particular ideologies or groups of people are presented within the narratives of games in relation to explicitly geopolitical issues (Halter 2006, Power 2007, Der Derian 2009, Huntemann and Payne 2010). For instance, Power discusses how games such as *America's Army* work to construct crude stereotypes of Arab and Muslim cultures through the way enemies are visually depicted on screen. Distinct from these accounts, developing the concepts of envelopes and envelope power allows us to analyse how power operates on a 'non-representational' (Harrison 2000, McCormack 2003, Romanillos 2008, Carter-White 2009, Dewsbury 2009, Gallacher 2011, Anderson and Harrison 2012, Romanillos 2013) level of objects, forces and capacities to shape how bodies think and act. Investigating games through the concepts of envelopes and envelope power contributes to research in game studies, by showing how power works on a series of embodied habitual levels that do not necessarily, straightforwardly or only relate to the representational content of the objects in interface environments.

## Chapter outlines

The remaining chapters of the book build upon and further develop the preceding argument as follows. Chapter 2 examines the concept of interface and how it has been theorized in contemporary thought. Critiquing accounts that rely upon or articulate a definition of the interface as a meeting point between two distinct realms of being, I argue that interfaces should be understood as composed of inorganically organized objects. Developing the concept of inorganically organized objects through the work of Stiegler and Bogost allows us to avoid reducing objects in the interface to images or representations that are somehow less real than the extended objects we experience in everyday life. Understanding individual objects in the interface as having an autonomous reality is a necessary step to emphasize that objects have their own capacities and can be assembled to create interface environments that emanate what the book terms a 'resolution' and a 'technicity'.

Chapter 3 develops the concept of resolution to understand how objects and their properties and capacities are designed to appear as autonomous or homeostatic from other objects in an interface environment. Through a number of examples including *Uncharted 2* and *3* and *Battlefield 3*, the chapter demonstrates how designers attempt to create objects with an optimal resolution through modulating the relationship between high and low resolution. In doing so, I link the concept of resolution to neuropower to show how modulating the resolution of objects can generate synaptic relations in the brain and habits in the body that can create positive affects for the player. In turn, the chapter suggests that creating objects with particular resolutions is central to how space appears within interface environments. There is no 'space' in interface environments, only processes of spacing in which objects appear as distinct and differentiated from one another, according to their different resolutions and how these resolutions are arranged.

Alongside resolution, Chapter 4 argues that objects in interface environments also have a technicity. Unpacking and modifying Stiegler's account of psychopower and attention, I define technicity as the way that objects within interface environments operate to durably fix the now of perception. Through an examination of the hit box and aiming systems of *Modern Warfare* and the combo system in *Super Street Fighter IV*, the chapter identifies two logics of technicity: homogeneity and irreversibility. These logics work to shape temporal perception around the present moment by spatializing time into a series of static locations or 'now points' and, thus, form a key way that videogame interfaces construct psychopower.

Drawing together the concepts of resolution and technicity, Chapter 5 argues that videogames produce particular envelopes, or localized foldings of space-time that emerge from the resolutions and technicities of objects that make up an interface environment. These envelopes generate a form of envelope power that shapes how space and time appear as modes of potential for individuals that engage with them and, in doing so, modulate

# INTRODUCTION

the relationship between players' memory and anticipation. By examining the historical development of the interface systems in the *Final Fantasy*, *Resident Evil* and *Metal Gear* series, we can examine how their changing designs attempt to create different forms of envelope and with them different forms of 'presentness'. Specifically, the chapter argues that the videogames it discusses produce envelopes in which concerns around the past and future are marginalized in relation to the present. However, rather than being simply debilitating or disabling, these forms of presentness work as a form of envelope power to actually increase players capacities to sense difference, even though these capacities are ultimately cultivated to create economic value for the producers of these envelopes, rather than for the benefit of players.

Thinking more broadly about questions of human encounter and relation, Chapter 6 works to show how the envelopes generated by videogames have wider implications for the kinds of affective response that games enable and encourage. Developing the concept of ecotechnics, defined as the way in which technical objects separate out and put beings into contact with one another, the chapter argues that online statistical systems such as *Call of Duty Elite* and *Halo Waypoint* operate to hold bodies in opposition for long periods of time in order to generate a form of affective value in which sensorimotor skill is valued over all other forms of engagement. The types of sociality possible in videogames are then relative to the narrow envelopes these ancillary systems help to cultivate.

Chapter 7 draws together the insights developed across the six preceding chapters to think through the implications of videogames when they are understood as generating envelopes. The kinds of envelope produced by contemporary big-budget videogames are spreading to other objects and areas of life through processes of 'gamification' in which tasks and activities are framed as kinds of game in order to improve motivation and productivity. Examining technologies such as Google Glass, Nike Fuel Band and smart meters, I argue that the problem with these systems is how they attempt to generate envelopes that focus users' perception on a continuously modulating present tense at the expense of creative thinking in relation to future or past possibilities. Drawing upon the work of Stiegler and the concepts developed across the book, I present three ways that we can respond to these emerging forms of enveloped life and work to question and counter their potentially limiting tendencies.

The book closes by reflecting on how the concept of interface envelopes contributes to work on videogames, interfaces, objects and accounts of power under late capitalism. If interfaces are understood as complex sites of spatio-temporal emergence, this opens new ways of thinking about objects more generally. Rather than just inert tools or lumps of matter, interfaces of all kinds are central to the shaping of perception itself and, with it, how we anticipate, recollect and prepare for events in a world that is moulded by and emerges from the very logics of the interfaces we use to engage with this world.

# 2

# Interface

*[I]t is no coincidence that the most distinctive new place in the innovated medial world is the interface, which no longer refers to the space of encounter between faces, but rather the contact point between the face and the non-face, or between two non-faces.*

SLOTERDIJK (2011: 189–190)

*To touch something is to make contact with it even when remaining separate from it because the entities that touch do not fuse together. To touch is to caress a surface that belongs to something else, but never to master or consume it. It requires a certain space between beings, but also an interface where they meet.*

HARMAN, (2012: 98)

As Sloterdijk argues in *Spheres*, a key change that digital interfaces have brought about is a shift away from communication occurring between human faces to a situation in which bodies communicate via hands and eyes that engage with, and respond to, a variety of computationally driven surfaces. At the same time, Sloterdijk points to a core assumption that underlies a variety of accounts of the digital interface. Namely, that interfaces are a point of contact between separate types and categories of beings. For Jean-Luc Nancy (2000), on whom Harman is commenting, the trope of the interface does not simply point to a pre-existing separation between discrete beings, but actually shows that interfaces are a site of differentiation where the distinction between things becomes sharply contrasted and apparent, through the very process of contact. These two diverging accounts point to the ambiguous quality of the interface, which can equally be considered a point that allows different beings to come into contact with one another as much as a point through which differences and distinctions between things come to be measured and qualified.

This dual movement between separation and contact is repeated throughout various accounts of the digital interface. For example, as Bolter (2007: 198) puts it: 'the interface is the point of contact between the artifact and the world'. Or as Cramer and Fuller (2008: 150) suggest 'interfaces are the point of juncture between different bodies, hardware, software, users and what they connect to or are part of'. This logic of connection and separation is just as strong in psychoanalytically inspired accounts of media. In 'Interface Fantasy' for example, Nusselder (2009: 4) argues that: 'the interface is the gate leading humans into cyberspace, connecting us to the matrix, while simultaneously, because of its particular formations, still separating us from it as a whole'.

Rather than understanding interfaces as the outcome of a process that connects different domains together to complete some task or as images or symbolic representations, I suggest that interfaces should be understood first and foremost as environments of inorganically organized objects, which communicate through processes of transduction. To make this argument, the chapter is organized into three main sections. The first section offers an overview of current interface theories and identifies the key logics that underlie them. The second section defines and discusses what is meant by the term 'object', how interfaces can be understood as sets of inorganically organized objects and how these objects communicate with one another. Through this analysis, the third section suggests interfaces are best understood as environments or ecologies of objects. Analysing key examples from the videogame engine Unity, I show how the objects that make up these environments not only communicate with one another, but also communicate with the player. Understanding how interfaces operate as environments is key to the following chapters, where I specifically argue that games designers modulate the resolution and technicity of the objects that make up these environments to create interface envelopes.

## Interface theory

Within work on technology and new media studies, at least five distinct approaches to the interface can be identified. Interfaces are understood as (i) computational devices, (ii) a set of human practices, (iii) a medium for transmitting cultural logics, (iv) a link between analogue and digital and

(v) affective surfaces. By unpacking these five approaches, a common issue can be identified with all of them. Namely, that they position the interface as a bridge between two distinct modes of being (even if they don't always distinguish between the same modes of being), and in doing so reiterate an instrumental logic that reduces the interface to tools used to complete human tasks. Examining these accounts in detail is important in order to show that the way they theorize what these different realms of being are shapes assumptions about how the processes and mechanisms that connect them together work.

## *Interface as computational device*

Firstly and perhaps most obviously, digital interfaces have been understood as a set of computational devices. In *Software Studies: A Lexicon*, Cramer and Fuller identify five levels of interface within a computational system. These are:

> 1. Hardware that connects users to hardware: typically input/ output devices such as keyboards; 2. Hardware that connects hardware to hardware, such as network connecters; 3. Software or hardware-embedded logic, that connects hardware to software; 4. Specifications and protocols that determine relations between software and software; 5. Symbolic handles which...make software accessible to users; that is 'user interfaces', often mistaken in media studies for interfaces as a whole. (Cramer and Fuller 2008: 149)

Here, interfaces are both experiential objects and symbolic systems. The interface in a PC is both the graphical design of the operating system and the underlying code that manages this system. At the same time, one can access the graphical interface only through the physical interface of the mouse and keyboard. This chimes with Galloway's theorization of the interface as a set of intrafaces. Here, 'an interface is not a thing, an interface is always an effect. It is always a process or translation' (Galloway 2013a: 33). Galloway goes on to suggest that a threshold theory of the digital interface, in which the interface is understood as a kind of door or window into another domain, is unhelpful as it covers over the complex processes that enable particular entities to appear as part of, yet distinct from the digital interface of which they are a part (also see Rose and Tolia-Kelly 2012, Rose et al. 2014).

One could productively use this account of the interface to understand videogames. Any videogame utilizes all five definitions of the term interface that Cramer and Fuller outline. All forms of videogame involve some connection between hardware and user, be it through control pads or gestural recognition cameras. All videogames also involve connections between hardware and hardware, be it wirelessly through a dualshock controller to a Sony Playstation 4 or via wired USB ports in Microsofts gestural 'kinect' controller. Videogame consoles also utilize hardware-embedded logic that dictates how software designers can utilize and access the various processing units in the console. Furthermore, game engines that are used to design games have to interface with lower-level software and firmware in the console in order to operate. Finally, from the player's perspective, each game utilizes a graphical symbolic system that allows the player to interact and control what is happening in the game. Here, the interface is defined through a relationship with an outside as 'any process or structure that brings computational events into correlation with what is outside of them, including other computations' (Antic and Fuller 2011: 130). Inherent to this technical account is that an interface is a contact point between different types of entity, such as code, electricity, power units and so on. As a result, such accounts recognize that, upon investigation, any single digital interface quickly gives way to multiple interfaces between a variety of hardware and software that make up a particular device or object. However, these accounts tend to reduce the content that interfaces mediate to images or representations that are nothing more than the outcomes of the more solid 'real' technologies that enable these representations to exist. An avatar is only real in the sense that the player can manipulate it using a control pad, and a game environment is real only when it is visualized on a monitor and so on.

## *Interface as practice*

Second, digital interfaces have been understood as a set of practices. Farman (2012) sounds a note of caution against the purely computational understandings of the digital interface discussed above. He expands the concept of the interface beyond the technical to argue that particular devices themselves are not interfaces, and that any theorization of the interface requires a relational understanding of the human practices as well as the technical processes that coalesce and are formed around interfaces. In relation to mobile phones, he argues:

the mobile device is not an interface. Instead the device serves as part of the interface that is constituted as the larger set of social relations. The mobile device, on its own, cannot be considered an interface. The ways we use the device (i.e. the embodied practices of social space) as well as the ways the technologies of the device interact with other devices (by receiving calls, exchanging data, pulling electricity from a power source...) transform it into an interface. (Farman 2012: 64)

In relation to videogames, Taylor (2009) makes a similar point. She suggests that these interfaces might be best understood as assemblages which exceed the purely technical. In her words:

Games, and their play, are constituted by the interrelations between (to name just a few) technological systems and software (including the imagined player embedded in them), the material world (including our bodies at the keyboard), the online space of the game (if any), game genre, and its histories, the social worlds that infuse the game and situate us outside of it... (Taylor 2009: 332)

Both Taylor and Farman suggest that reducing the digital interface to a set of technical objects is problematic because it ignores how and why interfaces are actually used in practice. As a set of practices, interfaces can be understood on a symbolic and discursive level, as well as an embodied and habitual one. As Farman (2012: 63) puts it: 'the relations that define the interface are habituated.... The interface is produced through a sensory-inscribed body. The interface is deeply connected to our experiences of embodiment and how embodied practices get inscribed through cultural forces and habits'. For instance, when playing a game such as *Gears of War*, the player may well be aware of an enemy trying to kill their avatar, while being completely unaware of the underlying Artificial Intelligence (AI) rules that govern that enemy's movement or behaviour. Nevertheless, it is the AI algorithms that shape the habits of response the player uses to defeat the enemy. While the computational- and practice-based approaches to interfaces appear to begin from different starting points, they actually share the same underlying assumption: that the content presented by interfaces and the bodies that engage with this content are distinctive kinds of entity. Software needs hardware to operate, in the same way that an interface requires a practiced body to engage with and instruct that interface to complete tasks. In either case, interfaces operate as a point of contact that connects different types of object together, whether that object be human or non-human.

## *Interfaces as carriers of cultural logics and ideologies*

Third, developing the computational and symbolic understandings of the interface discussed above, a number of writers have suggested that interfaces can act to create, communicate and normalize principles about the way the world should be. In relation to videogames, Wark (2009: 20) argues: 'games are not representations of this world. They are more like allegories of a world made over as gamespace. They encode the abstract principles upon which decisions about the realness of this or that world are now decided'. In a similar vein, Manovich (2001) argues that the form and content of media cannot be separated from the way in which users relate to that media (also see Manovich 2013). In doing so, the interface:

> provides its own model of the world, its own logical system, or ideology.... For instance, a hierarchical file system assumes that the world can be organized in a logical multi-level hierarchy. In contrast, a hypertext model of the World Wide Web models the world as a non-hierarchical system ruled by metonymy. In short, far from being a transparent window into the data inside a computer, the interface bring with it strong messages of its own. (Manovich 2001: 76)

Unpacking the cultural logics described by Manovich further, Munster (2006: 21) has suggested that interfaces should be understood as forms of fold between body and world, based upon particular kinds of 'facialization' which create a 'command-control scenario by aligning human and computational cognitive processes or providing the machine with a "human face" to address our own'. That is, computers use a mode of address that draws upon the logic of human facial expression to encourage humans to conform to the computer's logic. Munster (2006: 21) sees these forms of facializations as extremely problematic because they operate to 'force information to be folded into our corporeality in predetermined ways that leave little room for negotiating other modes of interaction'. As such, not only does the interface operate purely as a surface of contact, but also serves to transmit a series of logics and assumptions that are internal to it. In a similar manner, Galloway (2013a: 75) suggests that ideology is not simply implicit to interfaces, but rather 'a certain networked relation is at play: software, the social and the act of interpretation combine in "an intense mimetic thicket" ... it is this thicket that can be called the political'. In this case, ideology is not explicitly transmitted through the design of an interface, but rather interfaces 'ask...a question to which the political interpretation is the only coherent answer' (Galloway 2013a: 75). For instance, the very action of using a search engine valorizes specific sites and individuals over others through ranking them and creates value from these rankings, without having to communicate any kind of formal ideology to the user. Yet, regardless of how ideology is defined or transmission theorized, this notion of interface is still based around the idea that interfaces operate as more or less transparent things that connect bodies and information together to complete some, albeit ideologically loaded, task.

## *Interfaces as link between digital and analogue*

Fourth, writers have theorized interfaces as a link or mediating object between the realms of the digital and the analogue. Here, the distinction between the embodied, cultural experiences of using interfaces and the computational materiality of the interface is made particularly explicit. Lunenfield (2000: xv) distinguishes between the digital and analogue in the following way: 'digital systems do not use continuously variable representational relationships. Instead, they translate all input into binary structures of 0's and 1's which can then be stored, transmitted or manipulated at the level of numbers or digits'. He gives the example of the difference between analogue and digital photography to make his point: 'The digital photograph, rather than being a series of tonally continuous pigmented dots, is instead composed of pixels, a grid of cells that have precise numerical attributes associated with them, a series of steps rather than a continuous slope' (Lunenfeld 2000: xv). Another way of distinguishing the analogue and digital is that the digital organizes information based on format, whereas the analogue organizes information based on medium. As Ernst (2013: 89) puts it: 'the whole distinction between analog and digital media art for the new archives rest in the fact that, in the technomathematical monomedium of the computer, it is no longer the material medium but rather the format that is the message'. Wark (2009) marks the distinction between the digital and analogue as a difference between how elements are counted. Whereas 'the analog is a variation along a line, a difference of more and less. The digital is divided by a line, a distinction between either/or' (Wark 2009: 84). In each of these definitions, the digital and analogue are conceptualized as different categories or realms of being that are distinct from one another.

Massumi expands on the relationship between the digital and analogue. Rather than just a technical distinction between different ways information is produced, stored or counted, Massumi argues that the digital can only create possible states based upon the data the computer has cut or split into discrete quantitative units, whereas the analogue is open and intensive, with the potential to create virtual possibilities. In Massumi's (2002: 141) words: '[t]he digital is ... exhaustively possibilistic. It can, as it turns out, potentialize, but only indirectly, through the experiential relays the reception of its outcomes sets in motion'. He goes on to describe the analogue as something that cannot be purely reduced to measurement through quantitative states or mathematical modelling. The analogue is 'a continuously variable impulse or momentum that can cross from one qualitatively different medium into another. Like electricity into sound waves. Or heat into pain. Or vision into imagination. Or noise in the ear into music in the heart.... Variable continuity across the qualitatively different: continuity of transformation' (Massumi 2002: 135).

According to Lunenfield (2000) and Massumi (2002), the analogue and digital are, in some respect, different realms of being that only meet, as Amoore (2013: 139) argues, when the text or image produced by the interface is 'read' by a human body: 'while the digital arrays of code in word processing or in digital photography remain bound by the programming of possibilities, the reading of the resulting text or image reintroduces the analog process and with it the potentializing relay of "fleeting vision-like sensation, inklings of sound, faint brushes of movement" (Massumi 2002: 139)'. This creates a situation, paradoxically, in which the digital is either reduced to a set of possible mathematical states or given a constitutive, productive power, compared to the analogue body, which can only read the data supplied to it. Wark (2009: 81) sums up this position succinctly when he states: 'the analog is now just a way of experiencing the digital. The decision on whether something can appear or not is digital'.

As such, this distinction between the analogue and the digital can become an unhelpful dichotomy. For example, in her discussion of subjectivity in a digital world, Elwell posits the digital and analogue as two separate realms of being that only meet or touch one another with the intervention of a human viewer or user. In her words: '[w]e already live in a feedback loop between the digital and the analog, where so much of what we do, believe, and feel is expressed, documented, and preserved in digital form' (Elwell 2014: 236). The problem with the distinction between the analogue and digital is that it reduces all objects within or produced by the interface to the same, ultimately undifferentiated, substance of 1's and 0's. As Wark (2009: 84) argues: 'in this digital cosmos, everything is of the same substance. Nothing is really qualitatively different. A cow, a car, your cousin: each has its shape and color, but in the end it's all the same, just stuff'. For Wark, particular objects within the screen-based image of an interface are reduced to the status of weightless representation or flat ephemeral image (on the relationship between image and representation also see Jay 1993, Rose 2001, 2003). A pointer in a PC-operating system or a gun in a first-person shooting videogame is not really a pointer or gun, they are just amalgamations of 1's and 0's that only exist as representations expressed through the 'hard' materialities of the technologies of the screen and hardware.

Although seductive, reducing the interface to a point of contact or separation between two seemingly separate realms of being is problematic. For example, in relation to the digital and analogue, the digital can end up being reduced to the quantified and quantitative, while the human is either reified as the site of an analogue potentiality or reduced to a passive user who can only read off the digital data. In this case, the digital is reduced to its capacity to make possible, whereas the analogue is the site at which this possibility can be converted into true potential as the human being experiences the text, and creates new associations in their mind, that could not be created by the digital computer alone.

## *Interfaces as affective*

Fifth, interfaces have been understood by writers such as Mark Hansen (2000, 2006b) as affective. Although there are many different accounts of affect (Gregg and Seigworth 2010), Hansen considers affect through a broadly Deleuzian lens as the capacity to affect and be affected by the world (Deleuze 1988, Massumi 2002, Thrift 2004). Affect then is the outcome of an encounter between various bodies, whether they be human or non-human, organic or inorganic. How objects and bodies affect one another is therefore based upon the capacities of those bodies, which are also subject to change through affective encounters. But, even Hansen's account of digital interfaces tends to reiterate a logic of the interface as a point of contact and separation between two different realms that are roughly divided between the human on the one hand and the technical on the other.

This is despite the fact that proponents of affect theory more broadly argue that affect allows an analysis of social life that can equally account for non-human objects in non-dualistic ways because it considers bodies in the broadest sense as 'not limited to the form of the human' (Anderson 2006: 735). Although many writers working with the concept of affect are keen to emphasize that affects can emerge and exist between non-human objects (Thrift 2005, Simpson 2009), Colebrook (2014) argues that accounts of affect tend to reduce affects to how they are experienced or made sense of by human bodies. In a similar way, affect is regularly discussed as independent of human will or consciousness, but these same accounts ultimately return to the body as the site that experiences or makes sense of affects. In Colebrook's (2014: 81) words: 'much of what passes as Deleuzian inflected theory champions precisely what Deleuze and Guattari ... [would wish to] ... go beyond'. As such, rather than study affects as tied to objects in and of themselves 'it is the event of privatization with forces or pure predicates being referred back to the single organizing living body that is celebrated by the "affective turn"' (Colebrook 2014: 81).

Colebrook's critique of affect theorists could be applied to Hansen's work on affect and interfaces. For example, Hansen (2004) argues that digital interfaces relate to one another and to the human bodies that engage with them through the medium of affect, while tending to conflate the relationship between affect (the capacity to affect and be affected) and affection. Hansen (2004: 135) defines affection as 'a particular modality of perception: an attenuated or short-circuited perception that ceases to yield an action, and instead brings forth an expression'. Affection is tied closely to human bodies in so far as it is understood as that which mediates between perception and action. In Hansen's (2004: 134) words: 'affection fills the interval between perception and action- the very interval that allows the body qua center of indetermination to delay reaction and thus organise unexpected responses'. At the same time, following Massumi (2002), Hansen (2004: 159) recognizes the autonomy of affect as something 'that passes through the body and can only be felt, often at a speed beyond and magnitude beneath the perceptual thresholds of the unaided human perceptual apparatus'.

In Hansen's (2004: 130) account of the interface, affect is neither in the image nor in the body, but emerges through the encounter between body and image: 'instead of a static dimension or element intrinsic to the image, affectivity becomes the very medium of interface with the image'. In this case, although Hansen (2004: 247) recognizes the autonomy of affect outside of any particular body, he ultimately considers how affect is expressed on the level of human embodiment to shape human capacities for sensorimotor action and anticipation. As Colebrook (2014: 89) suggests, in this kind of narrative, the force or autonomy of an affect becomes 'reduced to affection'. Or, in other words, the affect of the interface is reduced to how that affect is experienced by a human body that engages with an interface. This is problematic because it implicitly reiterates a humanist, instrumental perspective, which is precisely what most of the work on affect claims to be an attempt to move away from. In doing so, the specificity of objects that make up the interface becomes reduced to how they appear or are made sense of by humans who engage with that interface; their existence in and of themselves is often a secondary concern.

## Interfaces as objects

The five approaches unpacked above are certainly not meant to be exhaustive and they overlap in several ways. While these accounts begin from a variety of perspectives and examine different empirical instances of the interface, they are united by three logics that are common to all of them. The first logic is an understanding of the interface as a dynamic process rather than a static set of objects. The second logic is that interfaces are points of contact between different and distinct domains, often defined through the notions of the analogue and the digital or body and technology. The third is that interfaces are primarily instrumental in nature. That is to say interfaces exist and are engaged with to complete some human task. While undoubtedly helpful, these three logics tend to reduce objects in an interface system to the outcome of a human practice or computational process. In doing so, the objects themselves become black boxed tools, flat visual images, weightless representations or binary code that await activation, either by human beings who perceive or sense them or by the computer which runs code or follows a program. To some degree, the affective account of interfaces discussed above attempts to steer away from these logics by emphasizing the relation or encounter over and above the objects that make up the relation. Nonetheless, affective accounts still tend to prioritize the human body as the site that makes sense of affects, even if these affects are experienced on a non-representational, non-discursive or subconscious level.

Rather than understanding the interface as a series of connections between the human and technical, or body and machine or digital and analogue, we can understand the interface as a series of objects, the capacities and affects of which exist beyond a relation with the human. The benefit of this approach is that we can consider both how objects are designed by humans with particular intentions in mind, and how the same objects often exceed or confound these intentions. This allows us to take objects seriously while retaining a 'minimal humanism' (Thrift 2008) that recognizes the role that designers play in trying to achieve specific effects, even if they do not always fully succeed (Simonsen 2013, Ash 2014).

To rethink the interface as a set of non-instrumental technical objects, we can usefully draw upon what Levi Bryant terms a flat ontology. Bryant (2011: 246) suggests that a flat ontology refuses:

> to privilege the subject-object, human-world relation as either a) a form of metaphysical relation different in kind from other relations between objects, and that b) refuses to treat the subject-object relation as implicitly included

in every form of object-object relation. To be sure, flat ontology readily recognizes that humans have unique powers and capacities and that how humans relate to the world is a topic more than worthy of investigation, yet nothing about this establishes that humans must be included in every inter-object relation or that how humans relate to objects differs in kind from how other entities relate to objects.

A flat ontology would suggest that conceptually separating parts of the interface into their own distinct categories of being is unhelpful. Rather, the content contained 'within' or mediated by interfaces are objects that cannot be reduced to representation or image and are just as 'real' as the objects, such as screens or processors that enable content within the interface to appear. I want to claim the entities that make up the digital content of interfaces can be helpfully understood as what Stiegler (1998: 18) calls 'inorganically organised objects' or technics.

The term 'technics' refers to all manner of technologies that exist in a variety of different states. For example, an inorganic organized object could be a sound wave produced by a public address speaker, a liquid cola drink or a solid club hammer. In common with a flat ontology, Stiegler argues that technics have an existence outside of their relation with the human. By this, he means that any technical object is invented and evolves within a particular play of constraints or 'loose determinism' (Stiegler 1998: 34) that operates outside of the absolute control of the user or inventor of that object. Everything that appears within and frames the encounter with an interface can be considered an inorganically organized object that does not simply appear for or through human structures of access. For example, a weapon in an FPS game, or a pointer in a PC desktop system, is as equally real as the RAM or hard drive which enables the game or desktop to run, and as equally as real as the hard plastic controller or mouse a person uses to manipulate these objects. While a weapon in an FPS game may appear as an image, it is still an object that has been designed and created with computer software and has a capacity to generate qualities and affect the world beyond a relation with the player.

But, this is not to say that an account of inorganically organized objects simply recognizes that technical objects can exist as different states of matter (such as an image composed of pixels of light, or a gun shot existing as an organization of sound waves). Instead, inorganically organized objects are multiple, coexistent and singular at the same time. This might seem to be a strange and somewhat contradictory claim. To clarify, Bogost gives the example of the videogame *E.T.* The *E.T.* cartridge is, amongst other things,

'8 kilobytes of 6502 opcodes and operands, which can be viewed by human beings as a hex dump...a flow of RF modulations that results from user input...a type of integrated circuit...a molded plastic cartridge...a unit of intellectual property' (Bogost 2012: 17–18) and so on. From this perspective, 'there is no one real *E.T.*, be it the structure, characterization, and events of narrative, or the code that produces it, or the assemblage of cartridge-machine-player-market, or anything in between' (Bogost 2012: 19). It is all of these, very different, things simultaneously.

Therefore, inorganically organized objects only appear as single objects based upon the object's particular encounter with another object. In Bogost's (2012: 23) words: 'things are independent from their constituent parts while remaining dependent on them. An object is thus a weird structure that might refer to a middle-sized object such as a toaster as much as it might describe an enormous amorphous object like global transport logistics'. For example, an iPad appears as a single object to a human being who uses it, but is composed of multiple other objects that are present but not apparent to that user, such as the graphical processor or hard drive. At the same time, components of the iPad, such as its battery also only encounters the iPad as a single object, albeit a very different one. Rather than a tool used to access the internet or engage with apps, the battery encounters the iPad through the interface of the power convertor as an object that consumes its stored energy. In this way, objects are both one and many at the same time, appearing in different ways as different things, depending on what or who encounters it.

In summary, inorganically organized objects (and perhaps all objects) can be understood as a kind of autonomous, yet partial, 'unit' (Bogost 2008) that can only selectively encounter other objects. Drawing upon Harman (2011: 19), we could state that inorganically organized objects are 'defined by their autonomous reality...emerging as something over and above their pieces, while also partly withholding themselves from relations with other entities'.

This account of inorganically organized objects also questions a hard distinction between objective primary qualities on the one hand and subjective, human secondary qualities on the other. Drawing upon Harman's object-orientated ontology, Clough (2013: 123) explains:

> [a]...posture towards objects...turns attention toward the primary and secondary qualities of objects, given the latter, – color, taste, or heat are understood to be qualities of a subject's perception. In contrast, primary qualities belong to the object itself, such as length, width and depth. But in questioning correlationism and relationism, the distinctions between

primary and secondary qualities are troubled. It is suggested that all qualities might be considered as secondary, while being transformable in the relation objects have with each other, removed from the privilege of human consciousness.

Whereas technical objects are traditionally considered to relate to one another through their primary material qualities, humans use their cognitive and sensory capabilities to translate the primary qualities of things into subjective perception (secondary qualities). If we take Clough's reading of Harman seriously, however, we could argue that all objects 'perceive' one another in the sense that any object relation is a selective encounter and through that encounter particular qualities emerge that are not reducible to either objects properties. From this perspective, interfaces are sets of objects that continually encounter one another and generate particular qualities that are partially dependent on these encounters.

We can refer to this process of technical encounter in interfaces as transduction. Transduction has a variety of technical and philosophical meanings. In engineering, transduction refers to a process of 'convert[ing] one kind of energy into another kind of energy' (Morton 2013: 157), such as a 'record needle (magnetic cartridge) convert[ing] mechanical vibrations from vinyl into an electrical signal' (Morton 2013: 157). Within continental philosophy, 'transduction' is a term that is associated with Gilbert Simondon (Simondon 1995, Combes and LaMarre 2012, De Boever et al. 2012). For Simondon, transduction is a process 'in which activity gradually sets itself in motion, propagating within a given domain, by basing this propagation on structuration carried out in different zones of the domain … [whereby] … each region of the constituted structure serves as a constituting principle for the following one' (Simondon 1995: 30–31 in, Mackenzie 2002: 16). The usual example of transduction given by Simondon is the generation of a crystal in a solution, where the emergent structure of the crystal provides the basis for further crystallization, which in turn influences how the shape of the crystal develops (for other accounts of transduction inspired by Simondon, see Derrida and Stiegler 2002, Dodge and Kitchin 2005, Kitchin and Dodge 2011, Kinsley 2013).

For my purposes, transduction can be defined as the selective emergence of particular qualities via an encounter between technical objects with some end goal or effect in mind. In other words, transduction is a process by which objects in interfaces are organized by designers to produce particular qualities for other objects in that interface and for the people using that interface. The term 'transduction' is helpful because it offers a way of differentiating between random or accidental encounters between non-human-made objects that exist outside of the purview of human beings and an attempt to create and arrange technical objects in such a way to produce particular qualities that the designer or creator of those objects intends. As such, transduction is a two-way process. Neither object that enters into an encounter is transducer alone. Both objects are transduced and transducers in an encounter. The qualities that emerge from a transduction are thus partly determined by the interplay or relation between technical objects, but also partly determined by the qualities of technical objects themselves. The key point is that both object and encounter are organized in such a way to attempt to produce a particular quality that is desired by the designer.

## Interfaces as environments

In order to attend to this complexity of objects and their transductive relations, we can follow Bogost's (2012: 17) suggestion to develop a 'specific and open-ended' analysis of interfaces. Bogost (2012: 17) argues that this is necessary to 'make it less likely to fall into the trap of system operational overdetermination'. In other words, taking the reality of objects seriously means it is possible to analyse them without reducing them to a set of quantitative mathematical states on the one hand or subjective human experience on the other. This analysis would suggest that interfaces are best understood as environments made up of the arrangement of objects and the transductions that occur between these objects. To unpack and expand this account of interfaces as environments, we can turn to specific examples of objects in the videogame engine Unity.

Unity is a popular 'development engine for the creation of 2D and 3D games and interactive content' (Unity 2014: n.p.). The Unity engine employs a full graphical user interface that makes it easier for developers without large amounts of programming knowledge to create game content. Owing to its ease of use and the existence of a free version, Unity has become a popular games engine to create videogames on a variety of platforms, including Xbox One, Playstation 4, Wii U and others.

In Unity, an environment is constructed through creating and overlaying objects and providing them with components, such as position, rotation and scale. In order to create relations between these objects, developers can add a series of distinct layers or meshes. These meshes include a terrain mesh and a navigation mesh. Navigation mesh is a tool that is used to determine routes of travel for both the player's avatar and non-player characters that are controlled by the computer AI. The navigation mesh is laid over the top of the graphical mesh, which shows the objects that make up the environment, such as walls, slopes, bridges, doors and so on. Using options such as radius or height, the navigation mesh tool allows the designer to delimit where an avatar can and can't travel, by designating specific surfaces such as walls or other immovable objects as non-walkable. The navigation and graphical mesh are therefore separate objects that do not have to coincide with one another. For example, the navigation mesh could overlap with objects that appear in the graphical mesh, meaning the avatar could walk through seemingly solid objects such as boxes or cars. However, designers create coincidences between these layers to give the player a sense that the environment they are moving through is coherent.

In this case, each object in the Unity engine only selectively encounters other objects through transductions that disclose specific qualities. These qualities enable and disable specific capacities for the object in the engine and the user who is engaging with these objects. While the terrain built in Unity is as equally real as the avatar that is used to move through that terrain, the terrain has a more limited capacity to relate to other objects in the game environment than the avatar does. The terrain is constructed from polygons and pasted with multiple textures, which are created using other programs such as Photoshop, in order to appear as if it was constructed from a particular material such as stone, grass or gravel. For the avatar, whose movement is limited by the navigation mesh rather than the graphical mesh, the texture on the terrain mesh is invisible. However, for the player controlling the avatar the terrain texture may conjure associations as a particular type of terrain made from particular materials, which may in turn conjure associations about geographical locations where the terrain might be found.

The qualities transduced to the player through their encounter with the graphical mesh can then be further transduced into a variety of emotional states and responses. For example, at the intersection of the navigation mesh, graphical mesh and avatar, we could argue that, while a wall serves to limit the avatar's movement to a preset area and appears to the avatar as an absolute boundary, the wall can appear to the player that is controlling the avatar as a nuisance and create emotional responses of annoyance or upset because of its simple capacity to act as a limit to the avatar. While these encounters more or less occur at the same time, the same object can create transductions that produce very different qualities for the beings involved in that encounter (solidity for the avatar, annoyance for the user). In this case, no simple distinction between the objects in the environment and the interface with which the player interacts with these objects can be made. As Jørgensenn (2013: 106) puts it: 'it is not possible to separate the gameworld interface from the content because interacting with the gameworld interface *is* experiencing the game content'.

Interface environments are, then, assemblages of objects that are positioned and spaced in relation to one another in order to transduce qualities for both other objects in the interface and the user engaging with that interface. Using Unity as an example points to how we can analyse the interface as composed of inorganically organized objects without introducing a distinction between representation and reality on the one hand or image and world on the other. The key point here is that interfaces are not just a point of material contact or relation between different things, or that the content presented through interfaces are just images or representations. Rather than simply more-or-less efficient tools designed for human ends, interfaces are environments. However, this does not mean that interfaces simply represent particular kinds of environment, such as a desktop on a PC or a snowy landscape in a videogame (although they can certainly do this). Instead, I am arguing that interfaces are actual environments; they are ecologies of objects, each of which has their own reality and capacity for relation with other objects.

## Interface, object, transduction

Thinking about interfaces as compositions of objects, each with their own realities and capacities, rather than representations or symbolic systems, offers a different starting point both to develop concepts to understand what interfaces do to the bodies that engage with them and to develop a critique of these interfaces. Rather than the subjective 'individualization' of forces relative to a player or user's embodied biography when they encounter these interfaces, interfaces create qualities that are relayed and partially transmitted by the very objects that make up the interface environment itself. Interfaces do not just work on a discursive or embodied level to shape knowledge or beliefs about the world, but directly shape what bodies are and what they can do, through the qualities that emerge from the transductions between them (as we see in Chapters 3, 4 and 5).

The concept of inorganically organized objects is useful for thinking about how interfaces operate as sites of transduction. Thinking transductively cuts across theoretical distinctions between digital and analogue or body and screen to focus on the encounters and relations between objects. In the same regard, understanding all object relations as transductive can account for how non-human processes come to have tangibly human effects on the user of that interface. In other words, an object-centred approach to interfaces does not consider the interface as an encounter between different types of materiality, with the digital as a single object positioned on the one side and the world as a differentiated set of objects on the other. Instead, this approach suggests that we begin by understanding the objects that appear within an interface and seek to account for how they selectively relate to one another to produce particular qualities and environments.

This chapter has defined objects in interfaces as kinds of autonomous units that have a reality that is irreducible to the status of image or representation. This broad definition was necessary to rethink the interface as a set of objects rather than a point of contact between different material realms. As Chapters 3 and 4 go on to argue, objects in interfaces can be more specifically defined as having two elements, a resolution (or spatial appearance) and technicity (or temporal appearance). Resolution refers to the degree to which an object appears autonomous from other objects, and technicity refers to how objects work to fix the now of perception. Developing the concepts of resolution and technicity is key to understanding how interfaces produce envelopes. Envelopes are the ultimate aim of interface design, because a successfully produced envelope modulates both the qualities players experience and their attention, constituting what, in Chapter 5, I term a new form of envelope power.

# 3

# Resolution

> *[T]he object commands us to approach it by means of certain specific modes that bring it into optimal resolution: we find the right volume for music on a stereo, the right time of day for a swim in the river.... This entails the obvious fact that many conditions of observation exist for the same object, and that objects phase in and out of various levels of resolution, shrouded in a fog of accidents that remains present even when we do find an especially compelling mode of perceiving the thing.*
>
> HARMAN (2005: 182)

In *Guerilla Metaphysics*, Harman refers to a way of thinking about objects in general as having a 'resolution'. Harman argues that resolution refers to how an object encourages us to approach it, based upon the differing capacities or sensibilities of that object. Very loud music emanating from a stereo might cause a human listener to cover their ears or move away from the source of sound, whereas a rock would be unperturbed by the same sound waves. Furthermore, many objects have an optimal resolution (from the point of view of the object or person encountering them). This is the resolution that is most accommodating to a person or another object that meets an object, depending on the situation or context. For example, a golf club will have an optimal resolution for a particular play, depending on a variety of factors, such as the weight of the club, the height or weight of the person picking it up, the location from which the ball is to be hit by the club and so on. Similarly, an ice cube will have an optimal resolution that will allow it to exist for the greatest length of time as an ice cube, depending on the temperature of water that it is placed in. A high temperature will cause the ice cube to melt quickly, thus generating a less optimal resolution, whereas a low temperature will allow the ice cube to melt slowly, generating a more optimal resolution.

The concept of resolution is similar in some regard to Gibson's (1977) theory of affordances, while different in others. For Gibson, affordance is the relational outcome between two entities that shapes their potential for action (Ash 2013a). Gibson's account of affordance is predicated upon two objects with existing properties meeting one another, which shapes the kind of affordance that emerges. Different to this, objects as I understand them produce a resolution that is based on the qualities that are transduced through a specific encounter, which are dependent on what aspects of an entity meets another entity, rather than a totally present fixed object. From this perspective, objects are not simply static, but actively 'phase in and out of various levels of resolution' (Harman 2005: 182) depending on the particular aspect of the thing or person that is encountered.

Harman develops a concept of resolution to discuss any object at all. Nonetheless, his account is useful for understanding how videogame designers attempt to shape an object's mode of spatial autonomy. As I argue across the chapter, creating objects that appear autonomous and distinct, while also part of a broader environment, is a key goal in games design and ultimately contributes to the forms of power games designers attempt to generate. To make these claims, the rest of the chapter forms four main sections and a short conclusion. The first section develops Harman's concept of resolution to think about objects in interface environments and games design specifically. The second section links the concept of resolution to Neidich's theory of neuropower. The third section examines games designers attempt to generate objects with particular resolutions in order to construct neuropower. The fourth section then moves on to understand how the resolution of objects in games modulate across low and high states depending on the bodies and systems that relate to these objects. Through the examples of objects such as joystick interfaces and the software objects of multiplayer networking, the chapter shows how resolution and neuropower, while potentially powerful ways of shaping player action, are also precarious, and rely on very precise organizations of objects to be effective.

## Resolution

While Harman's account of resolution is a helpful starting point, he is not the only theorist or thinker to use the term to understand how objects appear in the world. For Jean-Luc Nancy, resolution is drawn from a musical vocabulary to think about the way sound vibrates through space. In music, vibration refers to a material process of expansion, radiation and transmission of sound waves. Vibration is a displacement and upsetting of air that meets and crosses the boundaries and thresholds that differentiate objects from one another. Nancy (2007) suggests the sympathetic vibrations of musical instruments work together to produce the instruments' (and the piece of music being played on it) resolution:

> what makes work is nothing but this: what makes it in its totality and as its whole is present nowhere but in its parts or elements...What we are calling work is much less the completed production than this very movement, which does not produce but opens and continually holds the work open – or more precisely maintains the work as this opening that it essentially is, all the way to its conclusion, even if this conclusion takes shape from what music calls *resolution*. (Nancy 2007: 27)

In other words, resolution is the sympathetic vibrations that work together to create a coherent note or chord for the listener. In the science of optics, resolution refers to the ability of an optical system to resolve detail in the object that is being imaged. A particular optical device such as a digital camera will have a resolution that is dependent on the lens and the type of sensor being used to translate the information it captures into data. In television sets, resolution refers to the maximum number of pixels a display can present. A high-definition 1080p set, for example, can display 1920 × 1080 pixels, totalling 20,73,600 individual pixels of information.

Whereas Harman (2005) and Nancy (2007) suggest that objects in everyday life have a resolution that is determined by any number of accidental encounters or factors, within digital interfaces, the specific mode of resolution of an object is carefully designed and tested in order to be as optimal as possible for the player. In a digital interface, resolution is the outcome of transductions between a variety of objects including screened images, software, hardware and physical input devices, all of which are centrally involved in the design of particular objects within an interface. Just like objects anywhere else, in digital interfaces, objects express their resolution through a variety of channels or qualities. Usually encountered via a screen or keyboard and speakers, these qualities include sound, colour, light and vibration. For example, a mouse pointer on a screen has a resolution that is shaped by, among other things, the pointer's software drivers, the resolution and type of screen upon which the mouse pointer is being displayed, the colour and design of the pointer, the cleanliness of the roller ball on the bottom of the mouse that tracks its movement and so on.

Drawing upon aspects of Harman and Nancy's reading of resolution alongside an optical definition of the term, we can begin to think about objects in videogames as having particular resolutions and consider how resolutions are brought into being by designers and players alike. In my account resolution means three things. First, resolution refers to the ability to determine what an object is and what it does through the way in which the object transduces particular qualities to the player's senses. Following the optical definition of resolution, the denser the amount of information that can be transduced to those senses by an object, the higher the resolution. Second, resolution refers to the ability for the player to control an object in a game through fine forms of sensory motor discrimination, which are enabled by different densities of transduced qualities. In this case, an object with low resolution would appear indeterminate and the player would be unsure of how to control or relate to the object. If an object in a game was of high resolution, then it would appear more determinate to the player. The player would feel as if the object had particular fixed properties, even though those properties were actually stable or durable qualities that were transduced in an encounter. An object with high resolution would also allow the player to develop a high degree of confidence in controlling the object with accuracy and speed. Developing Nancy's musical sense of the term, the resolution of objects in videogames has multiple qualities that consist of visual, audio and haptic information, which resonate through the body and contribute to whether the object is experienced at high or low resolution.

Third, resolution refers to the extent to which an object appears homeostatic from other objects in an interface environment. Homeostasis can be defined as a state of relative stability in which the parts of an object that make up a particular system, whether that be a single organism, a population of organisms or a technical object, tend towards equilibrium. Simondon argues that all technical objects have to exhibit some form of homeostasis in order to exist as such. Giving the example of a simple oil lamp, Simondon (1970) suggests: 'think of a lamp which would catch fire, which would not have this control allowing combustion to be stable. This lamp would not be destined to exist, precisely because it would be self destroying' (for a more detailed account of technical homeostasis, see Ash 2014). In relation to objects in interface environments, homeostasis refers to giving objects the appearance of a functional autonomy that is both linked to and separate from other objects in an interface environment.

Within this definition, high resolution does not simply refer to an object having a particularly high graphical resolution, and low resolution does not mean that an object is graphically simplistic. For example, a table in the graphically complex *Beyond: Two Souls* (2013) has a lower resolution than a table from the original, pixel based, *Metal Gear 2* (1987). This is because a table in *Beyond* is merely a prop that cannot be interacted with; it forms a barrier that the player can manoeuver around, but has no other qualities the player can encounter (such as movement). However, in *Metal Gear 2*, tables can be crawled under and used to hide from guards. Therefore, the table in *Metal Gear 2* has a higher resolution, because it appears to a greater degree as an autonomous object with its own set of qualities. These qualities also give the table a connection to other objects that make up the interface environment, such as the floor, which works alongside the table to create a 'space' through which the avatar can crawl. In both games, the tables are designed to create the appearance of a coherent world beyond the arbitrary and constructed nature of the games' level designs, but, whereas the table in *Beyond* always remains in the background, the tables in *Metal Gear 2* actively emerge as items that have a concrete relation to the player's actions and the rest of the objects that make up the interface environment.

As such, the resolution of digital objects in the interface is not fixed or essential to these objects but can actively modulate through a continuum between different states of high and low resolution. This process of modulation occurs at a number of sites: both within the design and execution of the software and hardware that transduces the qualities of digital objects, and the embodied techniques and habits users develop in response to these objects. Designers attempt to create certain objects and control schemes in games that maintain a high level of resolution for players, but design other objects purposely with a low resolution. In certain situations, designers attempt to modulate the resolution of objects between a high and low state to generate particular qualities for other objects in the interface environment and sensory qualities for the player. But players also develop a series of embodied techniques to respond to shifts in resolution that occur between themselves and the software and hardware objects of the game as they play.

# Neuropower

Constructing and modulating the resolution of objects in an interface environment can be understood as a form of neuropower. For Neidich (2002, 2010), neuropower can be defined as a process of shaping synaptic relations in the brain in order to prime consumers' thought and decision-making processes in ways that are beneficial to the companies that are working to shape those relations.[1] Neidich's account itself develops Lazzarato's (2006: 186) theory of noopolitics, understood as 'the ensemble of techniques of control that is exercised on the brain. It involves above all attention, and is aimed at the control of memory and its virtual power'. While noopolitics examines how attention is manipulated through present events (Terranova 2007, Munster 2011), neuropower focuses on how objects and events remap the brain over longer temporalities. For Neidich (2013: 228):

> Neuropower...is not about the modulation of the attentive networks in the real present cultural milieu. Instead, it is about the rerouting of the long-term memories into working memory according to gradients of intensive affective flows, energy sinks, phase transitions, basins of attraction and stochastic and random resonances. This is the key to its link to the performative conditions of labor in the new economy. The machinic intelligence is not in the apparatuses of production as they once existed in the assembly-line of factories. Rather, it is within us as contemplative circuits mimicking the flows of new labor.... Neuropower is not about the production of a real object. Instead, it is exerted through a modification in the neurosynaptologics of the brain. In cognitive capitalism, neuropower works to produce changes in the material logics of the brain by affecting the brain's neurons and synapses....

In particular, neuropower works to shape consumers' decision-making through influencing their working memory. For Neidich (2013: 226):

> the new focus of power is not only on the false reproduction of the past – analogous to manipulating an archive; the effects of power have moved to the reconstitution of the working memory, elaborated by the forebrain in implicit decision-making processes utilized to form a plan or make a product choice. In other words, the new territory of neuropower is not past memory but future memory.

---

[1] Neidich is not the only theorist to develop the concept of neuropower. Other theorists who use the term, but define it differently, include Dunagan and Isin. For Dunagan (2010: 59), neuropower is considered more broadly than Neidich, as an 'enabling logic of a governmentality that sees the regulation of cognition, sensation, attention, mood, and mental fitness as part of its purview and responsibility'. Whereas for Isin (2004: 224), neuropower is considered more specifically as a 'call...to adjust conduct not via calculating habits but soothing, appeasing, tranquillizing, and, above all, managing anxieties and insecurities'. When using the term throughout the rest of the book, I am referring to and developing Neidich's definition specifically.

As such, neuropower does not work to introduce false memories into people's consciousness in relation to a consumption experience they may have had in the past, but actively attempts to influence the decisions people make in the moment regarding particular product purchases. Here 'memory is not the end of a process of retrieval but just the beginning. Memories are accessed and utilized to build a plan for future decisions and action' (Neidich 2010: 539).

The concept of neuropower emerges in the context of neuroscientific claims that the brain is a plastic or 'neuroplastic' organ. Neuroplasticity refers to the way in which the brain is shaped by its material experience in the world through a process of 'synaptic efficacy':

> If a synapse belongs to a circuit in frequent use, it tends to grow in volume, its permeability increases, and its efficacy increases. Inversely a little used synapse tends to become less efficacious. The theory of synaptic efficacy thus allows us to explain the gradual molding of a brain under the influence of individual experience, to the point where it is possible for us, in principle, to account for the individual particularities of each brain. We are dealing here with a mechanism of individuation that makes each brain a unique object despite its adherence to a common model. (Jeannerod 2005: 63 in, Malabou 2009: 7)

Exercising neural pathways through thoughts and actions can strengthen them, while degenerative neurological diseases may cause, or result from, the degradation of such pathways and create conditions in which it is impossible for new pathways to develop. As Malabou (2013) explains, this does not mean that the brain is deterministically shaped from outside; plasticity refers to the ability of the brain not only to give form to itself but also to receive form from the outside. This means that, from a neuroplastic perspective, the brain cannot be thought outside of a particular cultural or historical milieu. This 'enactive approach' to experience posits that: 'cognitive structures and processes emerge from recurrent sensori-motor patterns of perception and action. Sensori-motor coupling between organism and environment modulates, but does not determine, the formation of endogenous, dynamic patterns of neural activity, which in turn inform sensori-motor coupling' (Thompson 2005: 407).

Neuropower, then, is not abstract, but fundamentally linked to the embodied practices and habits of thought and movement that people develop through their everyday lives. The key point is that habits or practices are not simply learnt or retained in the mind as a series of rules, codes or representational schema, but operate in a feedback loop, informing and being informed by the material structure of the body and brain. As Malabou

(2012: xiii) puts it: 'the structures and operations of the brain, far from being the glimmerless organic support of our light, are the only reason for processes of cognition and thought ... there is absolutely no justification for separating mind and brain'. Or in Stiegler's (2009a: 193) words: 'consciousness is not the product of a mind as a frame preceding its content, but as those content themselves'.

Neuropower works to actively and deliberately shape processes of synaptic generation, and inform non-conscious and automatic habits and practices under a system of what Boutang (2012) terms 'cognitive capitalism'. Cognitive capitalism refers to the latest development of capitalism that relies on innovation, and the codification of non-codified knowledge in which 'everything can be combined with something else for profit, as life itself is redefined' (Thrift 2011: viii). While Boutang (2012: 165) defines value within cognitive capitalism as a kind of 'pollination of social relations' in order to produce know-how for companies, under neuropower value is generated on a more micro-level through creating habitual associations between body and brain in relation to individual bodies and the consumption objects they engage with. Shaping the resolution of objects in interface environments is, then, a key technique to construct, manipulate and maintain neuropower by shaping habits and practices on a largely implicit, unconscious level. To unpack the distinction between high and low resolution in greater detail and show how resolution is designed into interface objects to produce neuropower, we can turn to a number of contemporary videogames.

## High and low resolution

### *High resolution*

Creating particular objects with high resolutions is crucially important to videogame designers. This is especially the case in FPS games where the weapon the avatar holds forms the main way in which the player can relate to the interface. As Poole (2010: n.p.) writes: 'weapons are simple semiotic chips in the poker game that exists between you and the challenges of the videogame environment.... Good weapon design depends ... on a subtle balance of aesthetic considerations – the way it sits in your ... hand, the rumble or bang or surgical whirr of its firing sonics, and the effect it has on enemy flesh'. The resolution of a weapon in the game *Battlefield 3*, for example, can be understood as the qualities it discloses as players engage with it, including the way the weapon is modelled and animated, the way the weapon sounds and the way the weapon provides haptic feedback through the control pad.

In *Battlefield 3*, weapons are based upon approximations of real-world weapons, such as the M16 assault rifle and the L96A sniper rifle. In turn, the reload animations for each weapon mirror the weapon they are based upon, creating the appearance that each weapon is composed of individual parts that function mechanically in relation to one another and that each round a player fires corresponds to an actual round in the weapon's magazine. In gameplay, this is expressed through the way the reload animation is context-specific to whether the magazine has any ammunition in it when the player presses the reload button. If there is still ammunition in the weapon, the animation will show the previous magazine being ejected and a new one inserted, after which the weapon is ready to fire. If there are no rounds left in the weapon, the animation will show the new magazine being inserted, followed by an animation of the avatar's hand pulling a new round into the weapons chamber, thus making it ready to fire.

While seemingly just visual, these two animations take different lengths of time and, therefore, directly feed back into the gameplay experience and the decision-making process of the player. For instance, skilled players may reload when there are still rounds in the magazine to avoid the increased time it takes to reload the weapon from empty, which would make them vulnerable from enemy fire. How weapons are animated in *Battlefield 3* is a simple way to make them appear homeostatic, but also serves to generate neuropower by connecting the animation of the weapons to the embodied and synaesthetic experience of the player as they make strategic decisions about when and how to reload, based upon how many rounds the weapons have in a magazine and the length of time it takes to reload.

Another way in which the weapons in *Battlefield 3* emanate a high resolution is how the sound for the weapons is recorded and played to the player as the weapons are fired. Stefan Strandberg, Audio Director for *Battlefield 3*, suggests that a key part of good weapons sound is that the weapon sounds as if it is being fired in the environment within which the game is taking place. In his words:

> We worked a lot with reflections[,] layers and identity of place. We expanded on the way the weapons sound...in different environments going though urban, forests, canyons, open fields and indoor areas. It was key to build diversity on top of the identity of each weapon.... They share their footprint in the place they are fired, so in this way we could keep key signatures that built identity for specific weapons. The shared firing layers and reflections builds a believable homogenous lingering sound while the core weapon sound is there as a vital identity for players. (Misazam 2010: n.p.)

Recognizing that sound reverberates around space depending on the objects that make up that space, the sound of each weapon in *Battlefield 3* both reinforces that object's homeostasis and the homeostasis of the objects that make up the broader environment. This is achieved through each weapon having a core sound that is the same regardless of what environment it is part of, as well as being affected by the objects in the environment, that change depending on the location of the avatar who is holding the weapon.

As well as trying to create a sound space in which both weapon and environment appear to be part of, while distinct from one another, generating high resolution in *Battlefield 3* also involves dealing with issues of repetitions of sound:

> One of the first things we noticed in the internal multiplayer tests during production is that repetition and the patterns of iconic sounds are completely devastating to a believable soundscape. A gun shot might sound good when you design it and play it back on its own, but together with 50 other weapons and fired thousands and thousands of times you have to start thinking about all guns at the same time, and be very careful to treat them as individuals. All the weapons have to become one but still have identity, and they need to sit in the world. (Misazam 2010: n.p.)

Understanding that the weapons in the game would be heard alongside one another a lot of the time, the sound team worked to create a space in the frequency range of sound for each weapon so that they would not overly clash with one another. In this sense, creating high-resolution objects in interface environments is not about attempting to appeal to some naturalism or realism in which objects in the interface appear as they might do in real life. Rather, creating high-resolution objects involves designing entities that transduce qualities in ways that make the object appear to have a high degree of homeostasis in themselves, as well as in relation to other objects that make up the interface environment.

As I argued in Chapter 2, objects are transduced through encounters with one another to generate qualities, but the qualities that are transduced are not determined by or reducible to those encounters alone. In this way, the resolution of objects in an interface environment comes to be experienced by players as high or low, through their bodily senses and capacities. These feedback loops occur both within the body and between body and world. On one level, the body experiences qualities internally as a process of auto-affection or what Malabou (2012: 42) terms a form of 'primordial self-touching: ... [in which] ... the subject smells itself, speaks to itself ... [and] ... it is this contact that produces the difference of the self from itself, without which, there would be no identity and no permanence'. In turn, auto-affection can create new associations among different senses, thoughts and memories. These associations can take many forms, such as synaesthesia or what Deleuze (2005), following Marcel Proust (see Russell 1993), terms 'involuntary memory'. Synaesthesia is the mixing of different sensory inputs with different sense organs. For example, people with extreme forms of syanesthesia can taste colour or see sound. Involuntary memory involves the coupling of 'two sensations that existed at different levels of the body, and that seize... each other like two wrestlers, the present sensation and the past sensation, in order to make something appear that was irreducible to them' (Deleuze 2005: 47). For instance, involuntary memories might entail a strange sense of nostalgia generated as a result of a sound or smell recalled from a different part of an individual's life.

High-resolution objects can work on the level of synaesthesia and auto-affection to link the qualities transduced by inorganically organized objects to particular bodily sensations and states to create experiences the designer intends. For example, the animation of weapons in *Battlefield 3* can create haptic sensations of touching metal in so far as the sound of the reload and the texture of the weapon draw upon players' previous experiences of feeling metal surfaces or machines. In turn, these synaesthetic relations help the player feel closer to the weapon and thus more connected to the interface environment as a whole (also see Ash 2009). Resolution matters, then, because the resolution of objects shape the kinds of auto-affective and synaesthetic relations that come to be formed between body and brain and thus structure the relationship between body and world. Creating particular associations between body, brain and interface environment is a form of neuropower because, if successful, it can positively inform players' opinions about a game and how enjoyable it is, which is key to generating critical acclaim, positive word of mouth and, in turn, increased profits for the publisher of that game.

## *Low resolution*

Neuropower relies upon different objects achieving different resolutions. Indeed, it is important to emphasize that high and low resolution should not be understood as a dichotomy, with high-resolution objects considered as good game design and low resolution objects considered as bad game design. Games designers regularly create a number of objects in interface environments that are intentionally low resolution. In *BattleField 3*, the grass on the ground appears in low resolution as more or less homogenized lumps that cannot be affected by the player. This is intentional, both for technical reasons and to keep the player concentrated on the particular task at hand. If the grass appeared as high resolution, with each strand being independent of the other and connected to a root system, this might prove distracting to the player as they spent time manipulating the grass rather than shooting enemies or moving towards a specific location.

But, that being said, objects that are designed to emanate a high resolution do not always actually do so. In spite of the developer's best efforts, objects that should appear in a high resolution can appear with a low resolution. In turn, these low-resolution objects can fail to create positively affective synaesthetic relations between body and interface and so have difficulty generating neuropower. An example of this can be seen in the controversy around the release of *Uncharted 3*. *Uncharted 3* is the third game in the highly successful *Uncharted* series developed by Naughty Dog for the Playstation 3. All three games have been highly rated by the professional games press, with particular mention made of the quality of the visuals and cinematic presentation of the games many cut scenes and scripted events. However, shortly after the release of *Uncharted 3*, complaints emerged online from players who were having trouble with the aiming system used to control the avatar's weapons. In all three *Uncharted* games, the targeting reticule is controlled using the left analogue stick of the PS3 control pad. A popular gaming website shacknews reported: '*Uncharted 3* released two days ago, and many fans are crying foul. In spite of critical acclaim, many dislike the changes made to the controls of the single player campaign. Complaints include increased input lag and a general lack of responsiveness–something that's not present in the multiplayer component, and in *Uncharted 2*' (Yoon 2011: n.p.).

On gaming message boards, players had difficulty in describing exactly what the problem was with the controls (ZombiDeadZombi 2011). Fan made videos (Papercuts11 2011) attempted to measure the amount of delay between pushing the analogue stick and the character responding in game, with some commentators measuring up to 240 milliseconds of lag (szefu18 2011).

The developers Naughty Dog responded to fan feedback by giving an explanation about how the weapon mechanics worked in *Uncharted 3* compared to the earlier *Uncharted 2*:

> First off, the guns fire in a completely different way in *Uncharted 2* compared to what you're experiencing now in *Uncharted 3*. Keep that in mind, as it affects your perception of gun combat as a whole since it's very easy to want to compare the two sets of mechanics between the two games. In *Uncharted 2* the bullets would leave the barrel at a pre-set deviation when you were aimed in. What this means is that the bullets would not

fire straight out of the barrel all the time – they could come out at an angle. Therefore, you could have a target clearly in the reticle and still miss it by a wide margin. This was frustrating, because it was difficult to tell why you were missing a target.

As a result, we wanted to be sure you had a better grasp of whether you were hitting or missing a target. In *Uncharted 3* the bullets now fire straight out of the barrel 100% of the time. However, we have recoil – where the reticle moves/bounces as you fire. Therefore, it is easier to tell if you are missing or hitting a target. Now it is much more obvious when you are hitting or missing based on the reticle itself.

Aiming is identical to *Uncharted 2* – we took a look at the values side by side. We did adjust the sensitivity to be MUCH higher in *Uncharted 3* to give you a more precise feel. With *Uncharted 2* it was pretty much guaranteed you would aim in one of the 8 directions and it was hard to deviate from that (imagine it being almost like a traditional 8-way arcade stick). With *Uncharted 3*, you can deviate from the straight path from each of the 8 directions much easier and more precisely. (Richmond 2011: n.p.)

What is interesting about Naughty Dog's response to the low resolution of the weapons in *Uncharted 3* is the way in which the issues players describe actually emerge from an attempt to fix a previous issue with the ways the guns fired in *Uncharted 2*. As games director Justin Richmond suggests in the above quote, in *Uncharted 2* bullets fired from weapons did not travel in a straight line. This often created a sense of frustration for the user because perfectly aimed shots would not always hit their target. In *Uncharted 3*, bullets travelled in a straight line, but the sensitivity of the aiming reticule was increased and varying levels of recoil were added to each weapon. By changing the way in which the reticule moved, the designers increased the capacity for the weapon to be aimed more precisely. Although controlled using an analogue stick, in *Uncharted 2* the direction in which weapons can be aimed is limited in a quasi-digital sense to eight main points (up, down, left, right, diagonal left down, diagonal left up, diagonal right up and diagonal right down). In *Uncharted 3*, the movement of the weapon is more analogue, allowing players to move the weapon using finer haptic and spatial discriminations. This change was an attempt to increase the overall resolution of each weapon for the player but was ultimately unsuccessful. Rather than increasing the resolution of each weapon, the increased weapon sensitivity actually lowered the resolution for some players. This was because the designers challenged the expectations that experienced *Uncharted* players had formed about how the weapons operated.

As such, the controls in *Uncharted 3* broke down the pre-existing relations between body and brain that Naughty Dog had implicitly encouraged players to cultivate as they became skilled at playing previous games in the series. Changing the controls challenged the previous neuropower Naughty Dog had successfully cultivated in players' bodies and, in doing so, exposed the precarious nature of this neuropower. The neuropower constructed by Naughty Dog was precarious in the sense that players had difficulty in being able to adjust or alter the previous habitual relations between their bodies and brains to match the new resolution of the objects they encountered in *Uncharted 3*. In turn, this created a sense of disconnection, frustration and annoyance, which any interface designer is keen to avoid, precisely because these affective states make it harder for a player to enter into a captivated state while engaging with a game (Ash 2013b).

Changing the resolution of weapon objects across the *Uncharted* series suggests that creating a high-resolution object is not simply about shaping how players relate to and experience objects in one particular interface environment, but is also about creating durable neuroplastic and synaesthetic relations that work across and between different interface environments that form part of the same intellectual property. In the case of *Uncharted 3*, not only did the designers have to create weapon objects that fulfilled the somatic memories of players previously accumulated synaesthetic experience from *Uncharted 1* and *2*, they also have to create horizons of anticipation, in which the players' synaesthetic experiences are reaffirmed through the ways that future instalments of the game might feel in the player's hand. Designing objects with a particular resolution in mind is key to creating and maintaining a neuropower in the products games publishers and marketers attempt to sell. The weapons and control scheme in *Uncharted* are important, not simply as a means for the player to manoeuver the avatar around the interface environment, but because they create relations between the bodies and brains of players that take effort to construct and maintain.

From the perspective of neuropower, creating economic value and profit is about constructing products that cultivate particular synaptic relations in the brain. Here, value is exteriorized from the interface or object and held in a user's body and reactivated (or not) when the user returns to the interface. In turn, the player is rewarded (or not rewarded) by those somatic expectations being fulfilled by the resolution of objects in the environment. If the player is not rewarded, as the example of *Uncharted 3* shows, this can cause problems for the company creating this environment and impact upon both their reputation and profits. *Uncharted 3* points to the ways in which resolution is a relational achievement between the various objects that shape the qualities that are disclosed when using an interface. Indeed, it demonstrates that while neuropower 'is much more precise, powerful, and transformative than previous forms [of power]...it is also more ambiguous, uncertain, and fragile' (Dunagan 2010: 59).

## Phasing between resolutions

### *Software mechanisms*

While games designers attempt to create particular objects, such as weapons, with high resolution, in practice resolution is not an inherent property of a particular object in a videogame. Resolution is contingent upon a number of layers of software and middleware objects and other factors outside of designer's control, such as the network on which games are played (if they include an online component) and the input device that players use to control these games. As a result, the resolution of an object experienced by the player often shifts between particular phases of high and low resolution. Designers and players alike develop techniques to try and manage these shifts in resolution, both of which have different effects on shaping the player's capacity to anticipate and respond to what is going on in game.

For example, many online games (where users play in a shared gameworld, connected via the Internet) are plagued by what is usually referred to as lag – a temporal delay in the transmission and return of information from a player's computer (called the client) to the server on which the game is being played. For the user, lag is experienced as an asynchronicity between their inputs and the speed at which these inputs are expressed as particular forms of action on-screen. From a software perspective, lag is dependent on the type of networking architecture used to play online games. The most commonly used forms of networking architecture are peer-to-peer and client-server models. Peer-to-peer architectures are made up of various client computers that are playing a game. In a peer-to-peer system, any peer can act as the host for the game. Here, information about what each player is doing is shared directly between the players, one of which serves as the host for a particular game session. Client-server architecture involves a central server to which the individual players or clients connect. The Source Engine, developed by Valve and used in games such as *Half Life 2*, *Left 4 Dead* and *Team Fortress 2*, utilizes a client-server architecture. In this architecture:

> The client and server communicate with each other by sending small data packets at a high frequency (usually 20 to 30 packets per second).

A client receives the current world state from the server and generates video and audio output based on these updates. The client also samples data from input devices (keyboard, mouse, microphone, etc.) and sends these input samples back to the server for further processing. Clients only communicate with the game server and not between each other (like in a peer-to-peer application). In contrast with a single player game, a multiplayer game has to deal with a variety of new problems caused by packet-based communication. (Valve 2012b: n.p.)

As network bandwidth is limited, the server sends samples, or snapshots, of what is going on in the game back to the client. The server itself runs according to a particular tickrate – the number of times a second the server simulates the world (in Source engine this is thirty-three times a second) and sends snapshots of that world twenty times a second back to the client. When playing a game, the client player creates user commands such as firing a weapon or jumping (snapshots of the keyboard and mouse inputs) and sends them at the same tickrate that the server is running at back to the server. In the language developed here, a higher tickrate generates a higher potential resolution for players as the temporal distinction between individual samples sent to the client is minimized. This gives the player a feeling that the game is acting and responding smoothly to their input.

However, when individual packets of data are lost or delayed, this creates a sense of temporal dislocation for the player. While a weapon object, such as a pistol from *Battlefield 3*, may have a high resolution generated through its graphical design and animation in so far as it transduces qualities of solidity and weight to the player, this high resolution may be compromised by lag, which is created by latency. In a networked environment, latency may cause a weapon the player is using to shift resolution. For example, players may miss the target they were shooting at because the target has moved since the server last sent a snapshot of the world back to the client with the target's position. If latency is uneven throughout the game, this would be experienced as a shift from high to low resolution as the weapon phases from firing accurately to being inaccurate and unpredictable. At the same time, lag also threatens to compromise the construction of synaptic pathways in the brain that form the basis of neuropower. If the game does not reliably respond to the player's input, then it is impossible for the player to develop automatic and habitual techniques for playing the game.

As inadvertent phases in the resolution of objects can threaten the construction of neuropower, the Source engine employs a number of techniques to try and minimize these phases in resolution. For example, the Source engine employs lag compensation to minimize the phenomenological sense of temporal delay as the user plays. Lag compensation works through the server keeping a history of all recent player's positions for one second in order to make up for delays in transmitting information from the client to the server. The server applies a simple equation to estimate the time at which the command was entered on the client side:

> command execution time = current server time − packet latency − client view interpolation. (Valve 2012b: n.p.)

This means that the server takes into account the amount of time it takes for packets of information to travel from the client to the server and moves the avatar backwards to the position the player experiences on the client side of the gameplay. For example, if the player is shooting at an enemy that is moving from right to left in their field of vision, the speed of the enemy's movement means that by the time the server updates players' positions, sends this information to the client so that the client can then respond and their response is sent back to the server, the enemy may no longer be in the same position. Lag compensation moves all the players' positions back to the time the server estimates the command was inputted by the user, meaning the player hits the enemy in the position they perceived them to be on the client side of the game.

The resolution of particular objects in a game therefore phases between different states as they are mediated by and pass through a series of software objects. In the case of online multiplayer gaming, the network architecture is actively designed to try and maintain a high resolution between the various players and the environment in which the game is played. In other words, the lag compensation equation works as an object to maximize the potential for synaptic connections to be constructed through body and brain, by enabling the conditions for objects in the interface to appear in high resolution as homeostatic and autonomous.

The attempt to maintain a high resolution of objects in online games, and thus generate neuropower, drives an intense material economy beyond the games themselves. Multiplayer games that create contexts in which resolution is consistently low or regularly shifts from a high to a low state are widely condemned by players. Lag is a greater problem on peer-to-peer network architectures because the quality of the host's connection to the other clients in the network is not as stable as a centrally run professional client-server model. In turn, players are becoming increasingly sensitized to the differences between a peer-to-peer and client-server model, which is directly feeding into the networking and infrastructures games publishers are using for their most popular online titles (King 2013). For example, the console versions of *Battlefield 3*, Electronic Art's (EA) rival to the massive selling *Call of Duty* series, actively advertised that the game would run on dedicated EA servers, rather than a peer-to-peer model employed in the console versions of *Call of Duty* (Purchese 2011). The distinctions between different types of server architecture are not simply technical; these distinctions, and players' knowledge that it is the distinctions between these systems that cultivate more or less positive experiences, are now actively used to try to create competitive advantages for the companies creating these games and to differentiate their products from their competitors.

## *Player techniques*

Lag compensation is just one of many objects through which programmers and designers attempt to concretize the synaesthetic relationship between body and brain in order to generate an object's resolution and thus create neuropower. Alongside particular software mechanisms that are utilized to bring objects into high resolution for commercial reasons, players themselves also develop a series of bodily habits and other techniques to respond to the objects that make up the interfaces on which they play. An excellent example of this is the modification of arcade sticks and the associated techniques members of the videogame fighting community utilize in play.

Videogame arcade sticks emerged as a product in response to arcade videogames being converted to home videogames consoles. Arcade sticks mimic the set-up of the button layout and parts that are found on the arcade cabinet version of a game. In this way, they allow players to replicate the arcade set-up within the home to some extent. The rise of home-console arcade sticks has led to the development of a whole subculture that is concerned with modifying arcade sticks with different parts from different manufacturers, such as Sanwa and Semitsu. These parts include the type of stick gate that the player uses. The control sticks in arcade units have eight directional switches (up, down, left, right and diagonals) and the gate is a plastic plate that is screwed under the surface of the control stick. The gate acts as a boundary that limits the maximum movement of the metal stick shaft that players manipulate in order to activate the switches and move their characters around in game.

Stick gates are generally manufactured and sold in one of two shapes: square or octagonal, which alters how the stick feels to manipulate. Square gates provide more indeterminacy of movement compared to octagonal gates because there is not an associated corner to let the player know they have actively engaged the switch. The type of stick gate that the player chooses is dependent on the game and type of character they plan to play the game with. In the popular fighting game *Street Fighter IV*, square gates work better with 'charge' type characters that require the joystick to be held in one direction and then released, followed by a button press to perform a move. Octagonal gates on the other hand work better with 'shoto' characters whose moves are based on turning the joystick in quarter circles followed by a specific button press.

In this sense, gates can be understood as changing the overall resolution of the object of the arcade stick as a whole and the qualities they transduce. Using an octagonal gate closes down the space of lateral play enabled by the square gate and its four sharply contrasting sides and provokes and guides the stick along a series of smoother transitions between the straight lines that make up the shape of the octagonal gate. The effect of changing the gate allows users to perform specific moves more easily and with less sensitivity. This practice is widely known as 'riding the gate'. Rather than only pushing the stick the minimal distance required to activate the necessary directional switch, the octagonal gate provides an easier template and limiter to guide the players' movement of the stick in a quarter-circle motion.

Changing the object in this way produces a different resolution for the bodily senses involved in the action of turning the joystick in a quarter-circle motion. For example, the sense of touch and the sense of hearing are altered as is the relationship between the two senses. The octagonal gate means that players can push the stick until they feel it hit the limitation of the gate rather than use the sound of the stick switch clicking to let them know that they have successfully inputted a quarter-circle motion. The stick switch itself doesn't register any more of less movements regardless of whether a square or octagonal gate is used. Instead, it could be said that the gate generates a different resolution because the shaft of the stick now encounters an octagonal rather than round plate, and so transduces different qualities. The octagonal gate introduces more degrees of difference to be felt by the player compared to the square gate. In doing so, the resolution of the quarter-circle movement shifts, which in turn opens up other potentials for action, movement and sense to emerge.

The concept of resolution also allows us to further understand how the objects of stick and gate interact to generate and phase between different resolutions. When the player initially pushes the stick, the object of the stick appears in low resolution. This is because the degree of movement that the stick has before an input is registered creates a quality of indeterminacy in which the player finds it difficult to judge or assess how far they have pushed the stick and how far they need to push the stick in order to activate the directional switch. The stick gate itself could be understood as producing a high resolution when the stick comes into contact with it, by generating a quality of solidity. In doing so, the stick as an object becomes more determinate. It is easier to judge what direction the player is inputting and the limit of the sticks capacity for movement. In moving the stick in a particular direction and activating the switch for that direction, the object phases from low resolution to high resolution, for both the components of the stick (the gate and the stick shaft) and the player themselves.

These specific shifts in resolution are not neutral but actively open up and enable new kinds of sensory experience and action. For example, in response to the low resolution of the stick and its accompanying dead zone, users have developed a series of embodied techniques to combat this dead zone and bring the object of the stick into higher resolution. There are a number of different techniques for holding the stick component of an arcade stick. The specialist fighting game website eventhubs.com, for example, identifies six main stick grips. The main differences between the stick grips are what part of the palm is used to grip the ball top of the stick and how the fingers wrap around the metal shaft of the stick. Each of these grips could be understood as different ways of trying to solve the problem of the low resolution of the stick in relation to the stick gate and human operator and bring the stick into higher resolution for the player's senses.

A professional fighting game player, Daigo Umehara, uses a specific grip in which the smallest finger rests around the bottom of the stick while the other three fingers and thumb wrap around the ball top of the stick. In an interview with Japanese television station NHK, Daigo argues this grip is the best for certain fighting games, such as *Street Fighter IV* because of the way it allows him to control the execution of his moves (Ashcraft 2010). In the interview, Daigo shows how the stick grip helps him execute a special move that requires six directional inputs to be hit followed by a button press. These six inputs (down, diagonal forward and forward repeated twice) form two quarter-circle motions. Performing the move within the allotted temporal window requires that the inputs are executed exactly. If extra or unnecessary inputs are executed, the move will either fail to activate or will take longer to activate compared to inputting the move exactly. The specific grip that Diago uses enables him to increase the resolution of the stick and its capacities for movement in relation to his senses. The small finger that rests on the back of the shaft of the stick acts to limit the dead zone in which the stick moves without activating a switch. Moving his small finger to this area shifts the stick's resolution from low (where the dead zone of the stick is large and difficult to control) to high (in so far as the dead zone of the stick is now limited by the presence of his finger which makes it easier to judge the amount of space the stick has to be moved while still activating the necessary inputs).

The resolution of objects that make up an interface and how players respond to these resolutions form the key feedback loops through which players develop sensorimotor relations between body and brain. Interrogating these feedback loops, the examples developed across this chapter challenge Neidich's account of neuropower, which he suggests is essentially dulling, leading to less creatively able bodies. For Neidich (2014: 281–282), engaging with the objects of neuropower, such as computer screens and interfaces, leads to a 'diminution of time spent in deep thought, a loss of epistemological coordination in prognosticating, a decrease in the ability to analyse complex... situations leading to a supple and easily normalised subject'. Processes of neurological and cognitive normalization, enabled and encouraged by the objects of neuropower, literally create the conditions of possibility for what he terms 'psychopathologies'. As he suggests:

> I would like to consider the causes of the psychopathologies of cognitive capitalism, for example of panic disorders and attention deficit disorder, beyond those of overstimulation and feeble attention, and look at them instead as the leftovers of the incomplete process of the normalization of the plasticity of the brain and the imperfect instruction and superimposition of the real and imaginary upon the frontal lobes of the brain. (Neidich 2013: 222)

Here, Neidich links neuropower directly to disorders in which the subject or body can no longer cope with the stimulus of objects that attempt to habituate particular relations between body and brain. However, as the techniques players develop to use arcade sticks show, habituated changes brought about by neuropower are not just negative, dulling or psychopathological, nor do these habits create normalized forms of response. Rather, active engagement with these forms of interface encourages new forms of skill. Habit in the case of fighting game players is not simply about 'pure mechanism, routine process... [or a]... devitilization of sense', nor is it 'the disease of repetition that threatens the freshness of thought' (Malabou 2008: vii). Like plasticity, habit can be understood as a dynamic force that is 'both the site of change and movement, as well as incorporating the potential for bodily forms of fixity, continuity and stability' (Blackman 2013: 208). As such, habit is a productive process through which new skills are enabled and created. The different types of stick grip players develop to attempt to bring the stick into high resolution were not anticipated or planned by the designers of these games, but they do work to create a strong set of synaptic relations between body, brain and game and thus generate neuropower.

Modifying Neidich's definition of neuropower, we could suggest that neuropower is not about simply dulling or normalizing forms of habitual action or skill to a preset template laid out by the designer or producers of these technologies. Rather, neuropower also works to encourage creative forms of habitual modulation on the part of consumers. In this sense, neuropower is productive in that it encourages creative modes of response and skill that are not necessarily intended by designers, but still work to encourage and derive further neuropower from that skill and interface. In other words, it is important to emphasize that the creation of particular forms of embodied, habitual technique is not a subversion of, or challenge to, the logics of the interface envelope that videogames attempt to instil in players. This is because the techniques players develop in response to the resolution of objects in the interface environment and material interface primarily serve to draw the player more closely into a positive feedback loop with the game, which is the ultimate intention of the designers who create these games.

## Resolution, habit, power

Objects have a resolution or mode of appearance that dictates how discrete or homeostatic they appear within an interface environment. Games designers attempt to create specific objects with high resolution in order to generate strong synaptic connections between body, brain and interface and thus produce neuropower. Generating particular resolutions in objects in videogames is not a simple task. Any one object in a game is intimately linked to a broader ecology of other objects, which shape the capacities of that object. A gun in an FPS game may have a high resolution in single player mode, being animated and modelled to generate qualities of weight and responsiveness for the player. However, the same weapon may have a low resolution when used in an online match due to problems with the game's networking architecture or the client's connection to the network.

At the same time, designers can also intentionally create events or situations in which objects phase from high to low resolution in order to make the player feel powerless or out of control of what is going on. For example, in the climax to *Metal Gear Solid 4*, the main character 'Solid Snake' is forced to travel through a room of harmful microwaves. Through the majority of the game the avatar Solid Snake is experienced as a high-resolution object. The player can control Snake's movements in a variety of complex ways and also use a variety of objects to encounter and manipulate the environment. However, in the microwave scene, Snake becomes more and more wounded. At the beginning of the scene the player controls Snake as they have done for the majority of the game, by pushing the analogue stick in the direction they want Snake to travel. As Snake becomes wounded, the analogue stick stops responding and the on-screen instructions prompt the player to hit the triangle button repeatedly to force Snake to crawl forward. In this scene, control is actively taken away from the player, lowering the resolution at which the player experiences Snake as an object. Intentionally shifting the resolution of the Snake avatar serves to increase the dramatic tension of the scene and helps increase the player's affective engagement and relation with Snake as a character.

If designed correctly, intentionally modulating the resolution of objects between a variety of states is another technique for creating habitual relations between body, brain and game and producing neuropower. As such, it is important to recognize that generating neuropower is not simply about creating high-resolution objects. Successfully generating neuropower is a matter of modulating the resolution of objects when appropriate in order to respond to the contingency of player actions. As I go on to argue in Chapter 5, this process of modulation is what Neidich's account of power ignores, which is why I prefer to draw upon the concept of envelope power. If the resolution of objects in interfaces can be understood as their mode of spatial appearance, then objects also have a way of generating the appearance of time. In the next chapter, I consider how an object generates the appearance of time through the concept of technicity. Arguing that objects have both a resolution and a technicity is central to developing the concept of the interface envelope in Chapter 5.

# 4

# Technicity

*One of the dominant experiential effects of video games as a medium is the sense of agency induced by the player taking meaningful action, action that influences future events in the game. The very concepts of 'action', 'event', and 'influence' require an account of temporality in games – the myriad ways that temporal structure informs gameplay.*

ZAGAL ET AL. (2008: 846)

*Moving images suggest to human sensation both the now (whether here or there) in an ever unfolding and flowing present, and also that the present might be other than it is, that movement has the potential to shift the meaning of the present in unanticipated ways.*

HILLIS (2009: 174)

**Z**agel and Mateas argue that time in videogames can be understood through four specific temporal frames: 'real world', 'gameworld', 'coordination' and 'fictive'. Real world time refers to events that take place in the world around the player, such as the clock ticking or a car passing by. Gameworld time refers to times established by objects in the interface environment like the rising and setting of the sun in *Red Dead Redemption*. Coordination refers to events that coordinate the actions of multiple players (whether they are human or the artificial intelligence of the computer). For example, in the FPS game *Call of Duty: Modern Warfare 2*, rounds of free-for-all multiplayer matches are ten minutes long. Finally, fictive time refers to the narrative labels attached to specific sets of events. For example, a round in a game like *Civilization* may only last twenty minutes, but within the narrative of the game, the round may represent twenty years of progress. Time is then the relational outcome of sets of events rather a field or container in which events take place. Zagal and Matea's account is useful in distinguishing narrative time (as represented through in game events) and broader temporal frames that structure player experience without effacing the complex relations between the two.

Other accounts of temporality in videogames make similar distinctions between different forms of time in games. For example, in *Half Real*, Juul (2005: 143) discusses the differences between play time and fictional time, which is linked through projection: 'projection means that the player's time and actions are projected onto the gameworld where they take on fictional meaning'. As in Zagel and Matea's account of fictive time, this projection does not have to have a one-to-one relation with the fictional time of the game. While many action games are based on a real-time relation between player and game (such as *God of War*), Juul provides examples such as *Sim City 4*, where the player can set the speed at which time passes in the game.

While these accounts are helpful, they tend not to theorize or take into consideration the complex phenomenological and bodily processes through which time appears to human beings via the systems and processes that make up videogames. As Hillis (2009) suggests, any theory of temporality has to think through the temporal nature of perception itself and specifically how experiences of the present are structured around particular media objects. For Hillis, the unfolding of time necessarily brings with it a potential for the unknown and unexpected to occur, which folds into how the present is experienced.

Rather than focusing on the ways in which time is narrated or fictionally contextualized within games, this chapter examines how time is phenomenologically experienced by players and how time is shaped by the gameplay mechanics and software mechanisms of contemporary videogames. The first section develops the concept of technicity as a 'durable fixing of the now' to think about how gameplay and software objects operate to shape temporal perception. The second section links technicity to Stiegler's account of psychopower to emphasize the value that can be generated through shaping an experience of the now. The third and fourth sections identify two underlying logics that are common to the technicity of videogame design: homogenity and irreversibility. The technicity of objects in interface environments is designed around these two logics to maximize the psychopower generated. By examining the aiming and hit-box systems of the games *Street Fighter IV* and *Call of Duty*, I argue that the shaping of temporal perception works specifically to fix player's attention around a set of finely discriminated now points. The effect of this shaping is to concentrate perception on a continuously modulating present moment, creating a strongly captivating form of attention and, thus, new forms of psychopower. While the examples developed in this chapter are very specific, most games use hit-box technology and their gameplay is affected by the framerate at which they run. As such, the logics identified here are also relevant to understanding the technicity of objects in other interface enviroments and player techniques in videogames more broadly.

## Technicity

The concept of technicity has been theorized in a number of different ways by a series of thinkers such as Heidegger (2006), Simondon (2009) and Mackenzie (2002) (for an overview of these theories in detail, see Ash 2012b). Writing in 'Originary Technicity', Bradley and Armand suggest that the concept has its beginnings in Aristotle's account of *techne*:

> techne is an essentially inert, neutral tool whose status is entirely determined by the use to which it is put by human beings... techne is a prosthesis (pro-thesis, ie, an addition; what-is-placed-in-front-of) considered 'in relation to' nature, humanity or thought; one that can be utilised for good or ill depending upon who or what happens to wield it. (Bradley and Armand 2006: 10–11)

However, as Armand and Bradley argue, the concept of technicity, as developed by writers such as Derrida (1998) and Stiegler (1998), does not refer simply to technology as prosthesis. Instead, technicity refers to the originary nature of technology to human life:

> technicity names something which can no longer be seen as just a series of prostheses or technical artifacts which would be merely 'supplemental' (or supernumerary) to our nature but the basic and enabling condition of our life-world. From the watch we wear to the server we log into, we exist prosthetically, that is to say, by putting ourselves outside ourselves'. (Bradley and Armand 2006: 3)

Developing this notion further, technicity refers to the capacity for technology to give humans an orientation in time. Specifically, technicity can be defined by drawing upon Stiegler's reading of Heidegger's (1992) essay 'The Concept of Time' as a 'durable fixing of the now' (Stiegler 1998: 234). While Stiegler does not define technicity in this way, we can co-opt his reading of Heidegger to suggest that technicity means technology shapes how the 'now', as a phenomenological experience, emerges from a relationship between the memory of the past and the anticipation of the future, which is related to a particular activity or encounter between objects. As such, the 'now' of perception is not simply 'there' or apparent to human perception, but only 'appears' to humans through the sets of equipment and technology which make up the ecology of an environment. This concept of technicity complicates traditional phenomenological accounts of time consciousness and debates around the reality of the now for consciousness. As Dodgshon (2008) (following James 2011) argues, humans inhabit a 'specious present' in which we never experience a completely unextended now and we never experience a now in isolation. Or, in Pockett's (2003: 56) words, 'Now always shades on one side into the past and on the other side into the future'. This follows our common-sense intuition that 'all time what is past or in the future no less than what is present is experienced only through each now, an experience that leads us to tense time around the present' (Dodgshon 2008: 300).

Heidegger terms the way in which humans inhabit a present with a past 'behind' them and a future 'in front' a temporal ecstasis or horizon. This temporal structure is central to humans' ability to think, feel and exist. Heidegger (1962: 350) writes: 'the ecstatical unity of temporality – that is the unity of the outside of itself in the raptures of the future, of what has been and of the present – is the condition for the possibility that there can be an entity which exists as its there'. In other words, experience is given to humans by their temporal existence, which works to organize the potential for action.

For Heidegger, time can be discussed in three ways: as present, past and future. However, he opposes the common-sense notion of past, present and future as being three distinct realities, with the past behind us and the future in front. Instead, he argues that any temporal experience emerges through an orientation towards the future which allows the present to be experienced: 'Temporalizing does not signify that ecstases come in a succession. The future is not later than having been, and having been is not earlier than the present. Temporality temporalizes itself as a future which makes present in the process of having been' (Heidegger 1962: 350). In other words, humans look into and anticipate the future. This process of anticipation then works to structure an experience of the present. For example, waiting anxiously for a doctor's appointment, a patient may experience time to move very slowly.

Within this ecstatic structure, Heidegger argues that there are different modes of 'making present' or experiencing temporality. On the one hand, humans can encounter objects as *ready-to-hand*, as pieces of equipment that are used effortlessly and without thought to complete a particular task. This refers to 'a special kind of forgetting...essential for the temporality that is constitutive for letting something be involved. The Self must forget itself if, lost in a world of equipment, it is to be able to actually to go to work and manipulate something' (Heidegger 1962: 354).

Here, Heidegger refers to experiences in which individuals become absorbed in an activity and, in doing so, do not reflect upon or measure time in a quantitative way. For example, hammering nails requires concentration and, absorbed in such a task, minutes or hours may seem to disappear. On the other hand, humans can encounter objects in a *present-at-hand* way. This is experienced when individuals step back from their involvement in a situation and consider objects as distinct things, rather than part of a broader task. Using the example of a hammer, Heidegger (1962: 362) writes:

> in the physical assertion that the hammer is heavy we overlook not only the tool character of the entity we encounter, but also something that belongs to any ready-to-hand equipment: its place. Its place becomes a matter of indifference. This does not mean that what is present-at-hand loses its location altogether. Buts its place becomes a spatio-temporal position, a 'world point', which is no way distinguished from any other.

Looking at a hammer in terms of its weight de-contextualizes the hammer from the specific situation in which the hammer is being used. In doing so, the hammer becomes one of many hammers and an object that is relatively heavier or lighter than other objects. As such, the time and space in which the hammer exists becomes homogenized – a geometrical point that is no different from any other point. In this case, the ecstatic structure of temporality is experienced in different ways. In the *ready-to-hand* mode, time is experienced transparently as intimately bound to the object or activity in question. Whereas in the *present-at-hand* mode, time is experienced abstractly as a particular de-contextualized moment in which the object appears to the user or observer as a series of properties and qualities outside of a particular ecology of use.

Heidegger's ecstatic account of temporality has been accused of idealism. Harman (2005: 65) argues that Heidegger's separation of time into three modes such as past, present and future reduces time to the way in which it is experienced and appears to human beings. As a result, Harman suggests that Heidegger can't make any claims about the structure of time itself or how time appears to non-human beings. Stiegler's account of technics challenges this idealist tendency, while still recognizing the importance of Heidegger's distinction between time as experienced as either *ready-to-hand* or *present-at-hand*. Following my account of technicity as a 'durable fixing of the now', time is not reducible to the way it appears to consciousness because it is things themselves (as they exist outside of the human) that make time available. In other words, clocks, timetables, hammers and so on implicitly create an experience of the present and a way of relating past memory to future experience. This is implicit in Heidegger's (1962: 475) account. As he puts it:

> That which gets counted when one measures time concernfully, the 'now', gets co-understood in one's concern with the present-at-hand and the ready-to-hand.... Thus the 'nows' are in a certain manner, co-present-at-hand: that is, entities are encountered and so too is the 'now'. Although it is not said explicitly that the 'nows' are at the same time present as Things, they still get seen ontologically within the horizon of the idea of presence-at-hand.

In swinging a hammer to hit a nail, the user must be aware of why he or she is performing that action (perhaps to hang a picture on a wall). This implies a future outside of present experience. At the same time, the ability to swing the hammer relies upon past experience or memory. Stiegler (1998: 224) terms this kind of consciousness 'creative anticipation': 'Anticipation means the realization of a possibility not determined by a biological program' (1998: 151). In this way, the experience of a 'present' or 'now' of perception is constructed from an equipmental structure, through both the anticipation required to put this structure to use and the memory (both somatic and cognitive) to fulfil the task at hand.

Opposed to an account of time consciousness as a temporal succession, my reading of Stiegler and Heidegger posits that the present is secondary to a dual process of anticipation and memory which operate to set up how the present moment is experienced. This account of temporal consciousness chimes with Derrida's critique of Husserl (1991), on which Stiegler's work on temporal consciousness is based. For Derrida (1998), humans can never experience the pure now, for the simple reason that as soon as we think, write or say the now it has already passed. As Clarke (2011: 18) puts it: 'we are never able to intend this absolute now-point of perception as such – never able to be utterly present with the present – since...we can only grasp it as it slips into the past'. Or in Rushkoff's (2013: 6) words: 'the minute the now is apprehended, it has already passed'. In this case, the technicity of particular objects sets the limits around which something like a now comes to be experienced or known after the fact.

The technicity of objects works to shape an experience of time on both a *ready-to-hand* and *present-at-hand* level. When swinging a hammer without thinking about it, time is experienced corporeally as the movement of particular muscles and joints in the body and the relation between these movements. As a detached *present-at-hand* object, the hammer is experienced cognitively as a series of elements or traits that might be measured quantitatively through the weight of the hammer or its length. The 'now', therefore, has no objective existence and only exists for perception as structured around a dual process of anticipation and memory which actively emerges from an equipmental structure of technology.[1]

As Stiegler (1998: 222) argues, this process of technicity as fixing 'does not mean to determine but to establish. The tool of what is established is the vice that fixes the object of work, that makes possible both a determination and...the indetermination of the multiplicity of possible determinations'. How the now is established is highly contingent upon, and relative to, the technologies and practices within a specific locality. With this modified reading of technicity in mind, we could argue that technicity complicates the classic distinction between the existence of an objective time of the universe and the subjective time of human consciousness (Reichenbach 2012). Radicalizing Stiegler and Heidegger's account, we could argue that time is not some universal phenomenon or flow, nor is it only the result of human structures of consciousness. Rather, time is always localized to the relations between particular objects. From this perspective, a human using a hammer creates a now, but so does an apple falling from a tree and landing on the ground. There is, then, no absolute distinction between objective and subjective time, but rather a plurality of different localized times made manifest by the particular capacities of the bodies and objects involved in any given encounter. Indeed, this is close to how Bruno Latour (1988) understands temporality.

According to Harman (2009: 30–31): 'Latour holds that time is merely the result of negotiations among entities, not what makes these negotiations possible.... Or rather, "Time does not pass. Times are what is at stake between forces" (Latour 1988: 165).... In Latour's universe no external force, not even 'time', exceeds the full concrete deployment of actants'. Within this model, time is the outcome of the relationship between entities, or what Latour calls actants. As entities move about or their qualities appear to change, this also gives rise to the appearance that time is passing. The capacities of human consciousness shape this passing, in concert with the particular object

---

[1] This is not to say that contemporary time consciousness is entirely dependent on technology. As Glennie and Thrift (2002: 159) argue, time awareness was available before the advent of clocks or other technical objects, based upon 'a mixture of environmental cues (solar position, perceived qualities of light or dusk and so on) and the unequal (seasonally variable) hours of the...day'. As such, one cannot argue that technology is constitutive of contemporary time consciousness only that it actively and continuously shapes how the now is experienced.

being used by that human, which in turn form a nowness that emerges from the relationship between the users' previous memory about how to use the thing and the anticipation specific to what the thing is going to be used for.

As complex forms of interface device, videogames generate multiple temporalities as various software, middleware and hardware objects interact and relate to one another. All of these processes in turn shape the technicity that emerges when the player picks up the control pad and plays the game. How the technicity of technical objects in interface design sets the temporal limits around which a now appears is not an innocent process. Rather, it is highly mediated, political and contested. In this regard, shaping how the now appears can be understood as form of psychopower.

## Psychopower

Stiegler defines psychopower as the 'systematic organisation of the capture of attention made possible by the psychotechnologies that have developed with the radio (1920), with television (1950) and with digital technologies (1990), spreading all over the planet' (Stiegler 2006, as cited in Hutnyk 2012: 127). For Stiegler (2010a: 18), attention, which is the key currency of psychopower, is defined as:

> the flow of consciousness, which is temporal and, as such, is created initially by... 'primary' retentions – primary because they consist of apparent (present) objects whose shapes I retain as though they were themselves present. This retention... is then conditioned by secondary retentions, as the past of the attentive consciousness – as its 'experience'. Linking certain primary retentions with secondary retentions, consciousness projects protentions as anticipation. The constitution of attention results from accumulation of both primary and secondary retentions, and the projection of protentions as anticipation.

In other words, media industries work to capture attention by creating tertiary retentions (images, sounds, etc.), which then form the primary and secondary retention of viewers. In a media-saturated culture, tertiary retentions controlled by the programming industries come to dominate and inform consequent primary and secondary retentions. By shaping primary and secondary retention, psychopower works to de-chronologize time. Wark (2009: 82) argues that dechronologization refers to a process by which 'all of time becomes a series of discrete, equivalent and interchangable units. At each interval, time can be arrested and made to yield a number'. The result of this dechronologization for the human beings implicated in these technical systems is that 'there tends to be less consciousness of the past and there also tends to be less of a feeling for the future – and as such, there is an attenuation of the possibility of having an experience properly speaking' (Stiegler 2011a: 42). Focused on the here and now, individuals become unable to link what is happening in the present to pre-existing structures and processes that are responsible for bringing an event or object into being. By replacing secondary retention with tertiary retentions that are pre-constructed by the cultural industries, primary retention (perception) becomes 'standardised and...particularisable, meaning they are formalisable, calculable and finally controllable' (Stiegler 2010a: 99). Stiegler argues that this form of experience leads to a contemporary existence constituted by hypomnesis. Hypomnesis refers to tertiary retentions that replace the requirement for secondary retentions, such as numbers stored in a mobile phone address book. For Stiegler (2007: n.p.), the impact of hypomnesis is clear:

> We exteriorize in contemporary mnemotechnical equipment more and more cognitive functions, and correlatively we are losing more and more knowledge which is then delegated to equipment, but also to service industries which can network them, control them, formalize them, model them, and perhaps destroy them – for these knowledges, escaping our grasp, induce an 'obsolescence of the human', who finds itself more and more at a loss, and interiorly empty.

Hypomnesis is therefore a kind of enforced forgetting. In the case of mobile phones, many people cannot remember key phone numbers because they are used to them being stored as a tertiary retention on the phone itself. In doing so, these people become reliant on the phone and anxious about misplacing it and possibly losing the numbers and data it contains. To mitigate this anxiety, the phone company can then create and market additional services, such as cloud storage, which further increase the users' reliance on the phone and the technologies associated with it.

Stiegler's account of psychopower draws upon aspects of Foucault's theory of biopower and Deleuze's theory of the control society. In an interview with Crogan, Stiegler suggests that: 'retentions and protentions are always articulated or organized by what Foucault calls dispositifs, which I call retentional dispositifs...[or]...hypomnemata' (Crogan 2010: 165). Anderson succinctly defines biopower as the process through which 'life has become the "object-target" for specific techniques and technologies of power...[A]...mode of power based on the attempt to take control of life in general "with the body as one pole and the population as the other" (Foucault

2003: 253)' (Anderson 2012: 28). With a different emphasis from Foucault, Stiegler considers psychopower a product of the economic marketplace. For Stiegler (2010: 128), the question is less one of 'utilising the population for production than of establishing markets for consumption' and so 'the state's biopower is transformed into market psychopower'. In this regard, Stiegler sees psychopower as linked to what Deleuze (1992) called the control society, in which corporeal behaviour is modulated via non-repressive means. As Stiegler (2011a: 82) states: 'a control society does not only consist in the installation, throughout society, of social control, but rather penetrates into consciousness...and thus reinstates corporate control, not only by harnessing conscious time but by soliciting the unconscious through the channelling of conscious time, all of which is concretely expressed as a new stage of the grammatization of corporeal behaviour'.

Focusing on the way corporeal behaviour is organized by technical objects, Stiegler's account of psychopower emphasizes the role non-human objects play in the production and maintenance of power. As Lemke argues, Foucault focuses on humans as key to the production of biopower. In Lemke's (2014: 3) words: 'Foucault's work remains within the "traditional humanist orbit" (Barad 2007: 235), restricting agency to human subjects without taking into consideration the agential properties of non-human forces'. As I have been unpacking across the previous chapter and will continue to unpack through the rest of the book, recognizing and focusing on the agential forces of technical objects rather than human subjects results in a very different account of what power is and how it operates.

With these differences in emphasis in mind, it is important to note that my account of technicity is quite different from Stiegler's theory of attention, which underlies his account of psychopower. Whereas Stiegler argues attention, as a specific and general form of focus, is made possible by a relationship between present perception (primary retention) and past experience (secondary retention), I suggest that re-reading Stiegler's account of psychopower through the concept of technicity leads to a redefined notion of attention. In my account of attention, the now emerges from a relationship between protention and secondary and tertiary retentions rather than primary and secondary retentions. As such, psychopower is not simply about capturing or holding attention, but actively modulating attention within a set of perceptual limits organized around anticipation and memory. However, both accounts of attention agree that the now is technologically produced in the sense that the temporal flux of attention is inherently and constitutively structured by the technical environments in which humans find themselves. In videogame design, we can identify two key techniques that work to shape the appearance of the now and in turn generate psychopower: homogenization

# TECHNICITY

and irreversibility. In the next two sections, I develop a series of examples to flesh out this somewhat abstract theorization of technicity and its relation to psychopower through the analysis of a series of games including *Super Street Fighter IV* and *Call of Duty*. While the examples discussed in the following two sections are very specific, the mechanics and objects involved speak to techniques utilized in AAA videogames more broadly.

## Homogenization

The first logic of technicity in games design can be termed homogenization. Homogeneity refers to the spatialization of time into discrete and equal units of spatial measure. The example of *Super Street Fighter IV* can be used to unpack what this means and how homogenization generates psychopower. *Super Street Fighter IV* is the newest game in a long-running series of two-dimensional one-on-one fighting games. The first *Street Fighter* was released into the arcades in the 1980s and gave its developers, Capcom, a modest amount of success. The series became internationally popular with the release of the sequel, *Street Fighter II*, which was released in arcades in 1991 but then ported to a number of different home videogame consoles. *Street Fighter*'s success as a franchise is attributed to a number of distinct factors. Curran (2004: 30) suggests that *Street Fighter*'s appeal revolves around the visual aesthetic of the game:

> much of *Street Fighter*'s appeal lies in the spectacular nature of its combat; not just in the way it looks, but also in the way the player interacts with the game. Punches can be blocked and traded with kicks, leg sweeps leapt, and roundhouses ducked. Watching two good players act and react is like watching two skilled martial artists trade blows. It's the sort of skill that makes your jaw drop, makes you wish you were involved.

*Street Fighter II* was technically distinctive from other one-on-one fighting games on two accounts: the special move system and the combo system. What separated *Street Fighter II* from its competitors was the way in which the input movement of the special moves mimicked the characters' on-screen animation for the move. To perform a fireball with the character Ryu, the player moves the control stick a quarter circle forward followed by a punch button, which is translated into the character pushing his arms down and then forward so as to release the fireball. The other distinguishing factor was the inadvertent generation of the so-called 'combo' system, which is now a mainstay of fighting games in general. The combo system is a way players can link individual moves into complex combinations. If performed correctly, once the first hit of a combo move has been performed, the opponent will be unable to block following attacks. This combo system creates powerful techniques that provide competent players with a large advantage against their opponent. These two distinguishing features have been retained in later games within the *Street Fighter* franchise.

The combo system arises as a result of the ways in which the animation for the *Street Fighter* games is created. *Super Street Fighter IV* runs at sixty frames a second and each move in the game lasts a particular amount of frames. All moves are further split into three parts: startup, execution and recovery. Startup refers to the opening time of the animation before it actively hits an opponent. Execution is the amount of time in which the move is active and can damage an opponent, and recovery refers to the amount of time it takes before another move can be performed. If a move hits an opponent, it puts them into a hit stun state and, depending on the move, stops them blocking a following move for a particular amount of time. These periods of time are incredibly small. For example, the character Ryu's medium kick has a startup of three frames, an execution of five frames and a recovery time of sixteen frames. This means that the whole move is complete in less than half a second of real time.

As the *Street Fighter* series became more popular, players began to recognize and formalize how the move system operated and recorded what is known as 'frame data' for each character – lists of moves and their associated startup, active and recovery value. These lists provided a means of working out which moves could be linked together. By knowing the frame data for Ryu's crouching strong punch, players can look for a move with a startup of six frames or less and know that if it is performed correctly, this move will link and create a combo.[2]

There are two main types of combo in *Super Street Fighter IV*. Cancel combos refer to inputting one move, followed by another, which cancels the animation of the last move to perform the next. These are largely regarded as the easiest form of combo move as they do not require strict timing to perform. Link combos can create more damage, but are more

---

[2] Fans originally created frame data by video recording gameplay, slowing it down and counting the number of frames each move consisted of. Many of these lists are now available online at fan and community websites such as eventhubs.com and shoryuken.com. An example of fan-created frame data can be found here: http://www.eventhubs.com/guides/2008/nov/13/ryu-frame-data-street-fighter-4/. Capcom responded to the development of fan-made lists of frame data by producing its own, officially sanctioned frame data. For *Street Fighter IV*, these are available online at http://sfframedata.com/walkthrough/pages/ssf4_intro.php or in the game's official strategy guide.

difficult to perform because they require a more precise temporal window to input the following move in the recovery period of the animation for the preceding move.

Link combos require the most practice because of the small size of these windows. For example, the character Ryu has a number of link combos, including a jumping hard kick, followed by two crouching medium punches, followed by a crouching hard kick. This combo, which consists of four individual moves, has a number of links that make it difficult to perform. According to the frame data, a jumping kick has an execution of seven frames and no frames of recovery (because it completes when the character hits the ground). This is followed by a crouching medium punch that gives a hit advantage of five frames. The following medium punch has a startup of four frames, meaning that there is a window of two frames in which the second medium punch input must be entered (because the first active frame of an attack is the last frame of the startup animation). Too late and the opponent will stop being stunned and be able to block. Too soon and the move will not execute as the recovery animation from the previous punch will still be in the process of execution. The same is true of the next part of the combo. The crouching hard kick which follows the second medium punch has a startup time of five frames. The first active frame of an attack is the last frame of the startup animation. This means that Ryu's first active frame of hard kick is the fifth frame, the last frame of the opponent's hit state. This provides a window of one frame to make the link between the second medium punch and the final crouching hard kick.

The combo requires the player to develop an intense sensitivity to units of time that fall well below conscious awareness. The whole combo takes less than three seconds to complete and the player has to discern temporal windows within this period that are between 1/30th and 1/60th of a second in length. This sensitivity is developed in a number of ways. Players can learn through a process of trial or error or, as is common among advanced players, through using frame data and the game's training mode to contextualize their practice. In Stiegler's terms, the frame data works as a vice to separate out and de-contextualize parts of each move into individual components, which in turn allows players to practice and learn the move more effectively.

Returning to Heidegger's account of temporal ecstasis, we can examine how the link combos operate to shape and concentrate the temporal structure of player's action within the present moment. As Heidegger (1992: 17e) puts it:

> the clock shows us the now, but no clock ever shows the future or has ever shown the past. All measuring of time means bringing time into the 'how much'. If I determine by the clock the point at which a future event

will occur, then it is not the future that is meant; rather, what I determine is 'how long' I now have to wait until the now intended. The time made accessible by a clock is regarded as present.

Here, the numerical frame data acts as a clock, not by distinguishing the temporality of the combo itself, but through distinguishing 'how much' time there is between each individual component of each move and their relation to other moves in the combo. The frame data and the way the combo system operates serve to tense time around the present. In doing so, the technicity of the frame data and training stage creates a context in which players learn to respond to events with incredible speed and accuracy.

Recognizing the difficulty of performing combos and the development of frame data by players, Capcom added a specific training mode to *Super Street Fighter IV*, so players can work on their combos and moves in game. The training mode includes a specific training stage. Regular stages provide elaborate graphical backdrops for players to fight in, such as street scenes, cruise ships and jungles. The training stage drops these settings and instead splits the stage into a series of cubes, which are broken into four smaller cubes. Alongside these cubes, there is a thick red line running down the middle of the stage and one along the ground on which the characters stand. The effect of the training stage is to expose the geometrical nature of *Street Fighter's* stages and, in doing so, to provide a means of breaking down the space of the stage into precise and repeatable quantitative units (the individual squares which compose the backdrop and floor of the training room). This training mode is not simply about increasing players' skill so they can compete more effectively against other players, but is also about a training or honing of players' capacity to sense difference. This works to create more deeply attentive players and, thus, enables games designers and publishers to generate more value from the interface itself.

We can begin to think about how the now is constructed around the training stage. For instance, the background to the training stage can be understood as an intermediary that allows the player to link the spatial movement of the character with the time it takes for that move to be performed. It acts as a kind of clock through the way space is broken down into equal units of measure. Stiegler (1998: 212), quoting Heidegger (1992: 4E), suggests: 'the clock measures time (or change) by comparing the duration of an event "to identical sequences on the clock[,] and [it] can thereby be numerically determined"'. The static, unchanging background of the training stage acts like the face of a clock, providing a marker against which the speed and distance of different moves can be measured and compared. This has implications for how players approach particular situations in game.

For example, the character Ryu can perform a jumping 'tatsu' (spin kick) special move by turning the control stick a quarter circle backwards and then hitting a kick button. This move is usually performed from a standing position. However, if performed correctly, the move can also be performed while jumping on the spot, which gives the player a distinct advantage against their enemy. The timing to perform a jumping on the spot tatsu is much more difficult than performing a standard tatsu. Many advanced Ryu players use the training stage wall to aid the timing of the control input in order to complete the move. The move has to be performed at the maximum height of Ryu's jump animation. In regular stages, it is hard to tell where exactly this point is. In the training stage, it is clear that Ryu is at maximum height when there are three cubes below his feet. It is only then that the input should be performed, which in turn leaves enough time for the move to execute and potentially hit the opponent.

Through this process, the 'now' appears through a relation between the player's character, the animation system of the game, the player and the training stage. The training stage presents the now as an incredibly small unit of time through a fine spatial discrimination in the distance between the individual cubes that make up the training stage background. In training mode, the player becomes aware of the strict time limit that one has to enter the input commands to perform the tatsu. This spatialization of time allows the player to bring this process, which operates at the temporal limits of sense, to conscious awareness more easily than when the move is judged through the character's animation and the feel of the buttons and control stick alone. In Heidegger's language, the training stage operates to spatialize time producing an event in which the player can apprehend the temporality of the move in a *present-at-hand* manner. Through practice in the training stage, the technicity of the framerate of the game can be internalized until it becomes experienced in a *ready-to-hand* manner and repeated in other stages that do not have the cubes in the background.

Both the spatial structure and partitioning of the training stage and the frame data that has been developed to help players learn combo moves in the game are based on the homogenization of time into space. Heidegger (1992: 18e–19e) argues that the homogenization of time refers to:

> the tendency to expel all time from itself into a present. Time becomes fully mathematized, becomes the coordinate t alongside the spatial coordinates x, y, z. Time is irreversible.... Before and afterwards are not necessarily earlier and later, are not ways of temporality. In the arithmetic sequence for example, the 3 is before the 4, the 8 after the 7. Yet the 3 is not earlier than the 4 on this account. Numbers are not earlier or later, because they are not in time at all...

As Heidegger suggests, when de-contextualized from lived experience, specific numerical values do not themselves have a temporal sequence or temporality. Every value reduces time to an individual and comparable unit as a mode of the present. In *Super Street Fighter IV*, this homogenization of time occurs through the spatialization of moves and movement into frame values and spatial markers. These processes of spatialization serve to create a context in which players' temporal ecstatic structure is primed around the present. By creating very fine forms of spatial and temporal discrimination, the game operates to create a now that is based on an incredibly small unit of temporality and thus focuses players' attention in a very narrow temporal field of awareness.

The particular recovery values for each move and the startup values for potential follow-up moves, which the player can learn by consulting the frame data, operate to set the limits around which players experience the present and draw upon past and future within the three-second window of the combo. The player has to anticipate when to perform the combo and then perform the combo, which breaks up the three-second window into a smaller set of anticipations. They must also anticipate the opponent's next move. This anticipatory structure creates a situation in which decisions have to be continually made in a moment constructed by the objects in the game system. The potential and openness of the future and the sum of the past are reduced down into the relevant set of skills and decisions that need to be made in relation to the action in game. This homogenization of time does not determine the players' creative possibilities in game; it serves to organize the very temporal structure within which decisions are made.

The training stage habituates new capacities to sense difference, and enables players to explore these differences until they become unconscious and automatic performances that can be applied when appropriate. The technicity of the training stage not only attempts to construct attention through manipulating the now, but also works on the level of psychopower. As such, the technicity of the training stage does not simply hone a pre-existent form of attention, but actively constructs new forms of automatic memory, or routing long-term memory into working memory to create new forms of and capacities for attention. In Stiegler's (2010a) account of attention, the outcome of manipulating the technicity of objects is the construction of a new object from which to derive economic value. In contrast, manipulating the technicity of objects in the interface environment is about altering the relationship between body and brain. As we have seen in the previous chapters, this kind of neuropower is extremely valuable to games publishers. The videogame design in *Street Fighter IV* encourages the player to train their own capacities for action, which draws them further into the circuit of consumption because all of the effort and time that players put into getting good at the game means they are more likely to have loyalty to the *Street Fighter* series and therefore invest in future purchases in the franchise.

# Irreversibility

The second logic of technicity in games design can be termed irreversibility. Irreversibility refers to the ways in which games emphasize and encourage the user to concentrate on what appears to be a present now, rather than to look too far forward into future possible nows or into the past at previous nows. The logic of irreversibility can be unpacked through the example of *Call of Duty*, a popular military themed First Person Shooting (FPS) series of games. The series began in 1993 on the PC and at the time of writing is in its eighth iteration. The series has become increasingly popular since the release of *Call of Duty 4* on the Xbox 360, PS3 and PC in 2008. Part of *Call of Duty 4's* immense popularity has been attributed to its multiplayer mode in which up to sixteen players battle against one another in peer-to-peer networks connected through the Internet. Much like *Street Fighter*, statistics about the particular weapons in the game are now released in strategy guides and smartphone applications. These statistics unpack and demonstrate the variables that govern how the weapons respond within the game and their relative effectiveness in comparison to one another. For example, Cod4central.com lists each weapon's accuracy, damage, range, firerate and reload time amongst other factors. The availability of micro-level detail about the properties of each weapon allows players to experiment with how different weapon attachments (such as silencers or scopes, unlocked through completing challenges) and perks (statistical bonuses the player can unlock, such as faster reloading or higher health) alter and affect these variables and, thus, the players' chance of success within the game.

Knowledge about weapon variables also leads to the development of techniques that utilize the quirks of these objects. The 'no-scope' shot is one of these techniques. The no-scope is a technique which allows players to fire sniper rifles in the game without having to 'look' down the zoomed scope of the rifle and aim carefully.[3] This is understood to be advantageous because it

---

[3] The no-scope technique is widely attributed to the Xbox 360 user 'zzirGrizz', a veteran of the *Call of Duty* series of games, who began posting videos on YouTube showing his skills, in particular his ability to shoot enemies without zooming in from extremely long range, which allowed him to dominate his opponents. The original no-scope tutorial video can be viewed at: http://www.youtube.com/watch?v=DEoh3g1KqMk

allows players to react faster to what is happening within the game because they have eliminated the delay caused by the need to zoom in, aim and fire. The no-scope technique also allows users to fire without limiting their field of vision to what can be viewed through the zoomed scope (which can make them unaware of approaching enemies). The technique itself exploits a quirk in the game's hit box and aiming randomizer system. All modern FPS games utilize a hit-box system. The hit-box system is a series of cubes that surround the characters in the game, but which are invisible to the user. Hit boxes are used by the computer to calculate whether a shot that has been fired has hit an enemy or not. Hit boxes simplify complex forms of player geometry allowing the computer to calculate whether a shot has hit its target as quickly as possible (Valve 2012a: n.p.). In other words, it is easier for the computer to calculate whether a shot has hit an invisible series of rectangles rather than the multiple complex shapes of the visible character models.

In an attempt to avoid users being able to fire the sniper rifle with pinpoint accuracy without using the scope, the game designers implemented a randomizing effect to the weapon's accuracy. The aiming randomizer works to ensure that not all bullets fired hit the exact same spot when the weapon is not zoomed in. This means that players using the sniper rifle can't gain an unfair advantage by using the sniper rifle without zooming in. However, the no-scope technique enables users to counter this randomizing effect and to overcome – and even take advantage of – the pre-programmed aiming randomizer of the sniper rifle. In order to fire accurately without zooming in, users move the avatar, bring the enemy into the middle of the aiming reticules and then stop. If a user fires just as the aiming reticules stop, then the bullet will hit whatever is exactly in the centre of the reticules, which should (ideally) mean that the user can hit and 'kill' an enemy's avatar.

The no-scope technique involves the user learning the technicity of the aiming reticules and how they respond to player movement. Here, the now is fixed in a variety of ways. First, the now is spatialized through the individual lines that make up the reticules. The user has to be able to clearly see the individual lines at any one moment on the screen. Second, the now is fixed through the movement of the lines and the speed at which the individual lines travel between their inner and outer spatial limits. For Stiegler (2009a: 11), speed is a relational concept: 'speed is our experience of a difference in forces: speed in and of itself is nothing'. In the same way, the speed of the reticules emerge from the user's ability to recognize the outer and inner limits of the reticules' spatial distribution. What makes the no-scope technique so hard to perform is the narrow temporal window in which the player has to respond to the movement of the lines of the reticule as they travel from their extended maximum point to rest at their closed inner point. The movement of the lines of the reticule from extended to closed takes less than a second and the moment at which the lines come to rest is a fraction of this time. The way in which space and time are presented through the reticule fundamentally organizes the temporal structure of player action into a very narrow ecstasis. When attempting a no-scope, the player has to draw upon their memory of the maximum movement of the reticules and also be readily anticipating the split second at which they come to rest so they can fire. The reticule therefore operates as a technology of psychopower by shaping the relationship between memory and anticipation within a temporal window lasting less than a second. In other words, the now is structured within a very small window that emerges from the relationship between memory and anticipation that is, in turn, structured by the particular qualities of the aiming reticule.

The no-scope technique can also be understood in relation to Nedich's discussion of neuropower developed in the last chapter and in particular his account of future memory. Developing the skill to perform a no-scope shot relies upon previous memories being drawn upon in order to anticipate a future state, rather than a simple manipulation of the players' previous memories about the game or their experience playing the game. In this sense, the mode of attention or psychopower being produced by the no-scope technique is not a matter of inhibition or control, but is about generating and naturalizing new relations between body and brain. This technique is not an easy 'cheat' because it requires users to learn and naturalize the correct delay between the reticules stopping and the action of pulling the right trigger button on the Xbox control pad. This proves even more difficult to perform on moving targets. Being able to perform 'no scopes' consistently, then, is a highly revered skill in *Call of Duty 4*, which many users attempt, but only a (relatively) small number have mastered.

The development of these kinds of technique encouraged by the technicity of the objects in interface environments directly challenges Stiegler's argument that the attention economy is predicated upon the creation of passive or docile bodies. In Stiegler's (2010b: 30) words, mass media creates '[a] vast process of cognitive and affective proletarianization – and a vast process of the loss of knowledge(s): *savoir-faire* [situational know how], *savoir-vivre* [knowledge of how to live well], theoretical knowledge, in the absence of which all *savior* is lost'. However, as the examples of the training stage, combo and no-scope techniques, show, technologies of attention utilized in videogame design actively require and draw upon complex forms of skill and situational know-how in order to pull users more closely into the circuit of the attention economy.

While no-scoping is only available to expert users who have mastered the technique, the aim-assist functionality of the *Call of Duty 4* multiplayer engine also serves to fix the now for all players with the specific intention of capturing and retaining attention. The aim-assist functionality slows the movement of the players' weapon as the aiming reticules pass over an enemy avatar. The reticules slow down in order to give the player a greater opportunity to aim and fire at the target. The aim-assist system was implemented in the console versions of the game because of the difficulty of aiming accurately using the analogue sticks on the *Xbox 360* and *Playstation 3* control pads. The level of slowdown enabled by the aim-assist technology is commonly referred to as the 'stickiness' of the weapon, a term that refers to the feeling that the enemy player is coated in an invisible adhesive, which the weapon reticules cling to as they move through the hit box.

All weapons in *Call of Duty* stick to the hit box of an enemy player, but this stickiness is also affected by the sight employed. For instance, the holo sight provides more stick than the Red Dot Sight. Many players are unaware of either the aim-assist technology in *Call of Duty* or that different sights employ different levels of stickiness in relation to the enemy hit box. However, just because this form of stickiness is experienced in a *ready-to-hand* way does not mean that it does not have any effects on the temporal perception of players. One could argue that the aim-assist technology in the *Call of Duty 4* engine operates to shape the temporal ecstasis of player action. Turning the weapon to aim at an enemy, the auto-aim slows the weapon's movement as it passes over the enemy target. Through this process of slowing down, time is spatialized and fixed through differences in speed. The now is opened up by the slowing down of time through a process of spatial delay in which the weapon travels more slowly through the opponent's hit box. In doing so, the auto-aim serves to open the temporal window at which a player can hit their target. The auto-aim system serves to minimize the level of anticipation required to respond to the aiming event by lengthening the now point in which a vital decision is to be made. By opening the temporality of a decision-making event, the auto-aim system literally offers the player more time to react.

Here, designers have attempted to open the now for beginners as a way to cultivate and encourage their capacity to sense difference when they have little skill with the game or interface. This is important because when players are initially learning to play against human opponents online, they are more likely to lose repeatedly and so their potential to stop playing is higher. By introducing a subtle auto-aim feature, which many players are not even aware is in operation, game designers can attempt to mitigate against negative affective states such as frustration or anger that may quickly build when beginning to play online and so maintain player's attention. In different ways, both the no-scope and auto-aim systems serve to distribute players' sensory capacities amongst a series of technical systems. While the no-scope technique actively encourages a form of focused attention on a very narrow set of now points, and associated pasts and futures, the auto-aim system widens the now in order to minimize the difficulty of responding to an aiming event within a narrow field of possible nows.

In both cases, the technicity of these systems generates a temporal field of action that is governed by a logic of irreversibility. Heidegger (1992: 18e) refers to irreversibility as: 'what remains of futuricity as the fundamental phenomenon of time as *Dasein*. This way of viewing it looks away from the future towards the present, and from out of the present its view runs after time which flees into the past. The determination of time in its irreversibility is grounded in the fact that time was reversed beforehand'. In other words, irreversibility refers to the conceptualization of time as experienced as a present, with the present continually slipping into the past. Heidegger suggests that it is the throwness of *Dasein* into the future that sets up this temporal structure, which is covered over by the commonly accepted logic of temporal succession. The aim-assist technology opens the now at specific spatial points in order to emphasize this irreversible logic, in order to further fix temporal perception on the now. While *Super Street Fighter IV* uses a logic of homogenization to create equally spatialized nows, *Call of Duty 4* changes the length of the now in order to aid player performance depending on the situation. The designers of both games draw upon each logic of technicity in different ways, but do so for the same end: to fix the now around very small spatio-temporal discriminations for the explicit purpose of generating psychopower that is conducive to shaping and guiding player attention.

Heidegger (2006) argues, in a very different way, that technicity is a negative process in which space and time are reduced to quantitative forms of measure. However, technicity in videogame design is far more complex and ambivalent. Many gameplay and software objects sensitize players' bodies to develop what Stiegler (1998: 224) terms new forms of 'creative anticipation'. *Street Fighter IV* players have to practise for tens or even hundreds of hours to be able to perform complex link combos reliably in match or tournament play. Although these mechanics draw players' anticipatory capacities around a series of narrow now points, this is not to say that this necessarily reduces their capacity for anticipation. In the case of practising link combos or the no-scope technique, these mechanics actually encourage a more acute capacity for anticipation and creative response, albeit within a very narrow ecstasis of potential pasts and futures. As such, the psychopower of these attentional techniques does not result from inhibiting the players' capacity to think or act. Instead, these techniques productively draw upon players' capacity for creative anticipation, cultivating the conditions for new capacities to act to emerge. Rather than freeing the player from the attentional intentions of the game's design, these capacities actually draw the player into a more tightly controlled relation of attentiveness, in which the player's cognition is organized around a set of variables that are designed to produce economic value for the company making these games, above all else.

## Technicity, time, power

Game designers create objects and manipulate their technicity for the explicit purpose of creating psychopower and capturing and holding players' attention. The software mechanisms and gameplay mechanics that generate the technicity of the games discussed here are not limited to these games alone. Many contemporary videogames utilize these software mechanisms and gameplay mechanics. The majority of FPS games on consoles (such as *Battlefield 3*, *Halo 3* and *Team Fortress 2)* use aim-assist technology and the majority of modern fighting games (such as *Soul Caliber 5*, *Street Fighter X Tekken* and *Ultimate Marvel Vs Capcom*) use frame data. Furthermore, the majority of all videogames that involve colliding objects utilize hit-box technology, including titles as diverse as *World of Warcraft*, *Lego Star Wars* and *Batman Arkham Asylum*. More broadly, the two logics of technicity identified here (homogeneity and irreversibility) are not just limited to, or expressed through, these particular objects or games, but are better understood as logics that are common to many different modern videogames. In summary, the two logics of technicity discussed across this chapter are key to creating psychopower by capturing and holding attention. But, these logics are not fundamentally linked, nor are both logics present in every game. Some games utilize the logic of irreversibility, some the logic of homogenization and some exhibit aspects of both.

Reflecting on the points developed across the chapters so far, I have argued that interfaces cannot be reduced to an amalgamation of symbolic and representational software on the one hand and material hardware on the other. First and foremost, all entities that make up an interface, from particular objects in an image, to lines of code, are equally objects. As I have examined, these objects each have a particular resolution and technicity. These resolutions and technicities are not accidental, but carefully brought into being by a series of different actants, including hardware manufacturers, designers, programmers and others to create what Neidich calls neuropower (resolution) and Stiegler calls psychopower (technicity). To define and clarify the concepts of technicity and resolution, I have examined particular interface objects as more or less isolated or discrete things. However, as anyone who has used any interface system can attest, an interface consists of a variety of components working together. Objects in interfaces are carefully assembled, located and placed in relation to one another through processes of design, focus testing and so on.

Zooming out from the minutiae of particular objects in interface systems, we can now consider the cumulative effect objects have when they are assembled into what I have called an interface ecology or environment. Indeed, my main contention here is that, when successfully assembled together, objects in an interface environment have a potential power that is greater than the sum of the neuropower or psychopower of their individual parts. In the next chapter, I term the phenomena that emerges from a successfully assembled interface environment an *interface envelope* and the type of power these environments generate *envelope power*.

# 5

# Envelopes

In origami, the Japanese art of paper folding, an envelope can be constructed with only a single sheet of paper. A paper envelope covers, protects and surrounds what it holds and helps ensure its contents arrive safely in another time and space. However, at the same time, an envelope also partitions, separates and holds its contents apart from the various environments and elements external to it. While humans tend to associate the term 'envelope' with forms of folded paper or card, this is not to say that envelopes cannot be constructed from other states or forms of matter. In biology, an envelope can refer to the plasma membrane and cell wall of a bacterium. These components form a protective barrier that regulates what enters and exits the cell. They also provide the cell with mechanical strength (Silhavy et al. 2010). Envelopes can be constructed from solid biological matter, but envelopes can also be constructed from liquids, for example acids form envelopes in certain algae (Drews 1973). In geometry, an envelope refers to a curve that is tangent to each member of a group of curves (Bruce and Giblin 1984). On an XY plotted graph, the envelope is visualized as the boundary line that separates the plotted points from the rest of the graph. What unites these various forms of envelope is that in each case the envelope is a limit or boundary that creates a space through constructing a distinction between inside and outside.

Latour argues that envelopes can be understood as the outcome of a design process that seeks to fold environments and objects alongside bodies in order to allow those bodies to continue to exist in that environment. He gives the example of a space suit: 'When you check on your spacesuit before getting out of the space shuttle, you are radically cautious and cautiously radical...you are painfully aware of how precarious you are, and yet simultaneously, you are completely ready to artificially engineer and to design in obsessive detail what is necessary to survive' (Latour 2011: 158). This understanding of an envelope draws upon Sloterdijk's account of spheres. For Sloterdijk, spheres are 'air conditioning systems in whose construction and calibration, for those living in real coexistence, it is out of the question not to participate. The symbolic air conditioning of the shared space is the primal production of every society. Indeed humans create their own climate; not according to free choice, however, but under pre-existing, given and handed-down conditions' (2011: 46–48). Sloterdijk argues that spheres are the wide range of human-made objects and infrastructures that protect humans. Spheres can be buildings, literal air-conditioning units, life jackets, central heating systems, umbrellas and all manner of other things that allow humans to survive, communicate and exist alongside one another.

An envelope, then, is a carefully designed object that covers, protects or immunizes its user from some set of more or less dangerous or unwanted entities or forces while, at the same time, affording the user new capacities to act (such as allowing them to travel in hostile environments). Inverting Latour's sense of the term, an interface envelope can be understood as an emergent effect of a series of objects in an environment. Here the envelope is not used to cover, protect or immunize the user from its surrounding conditions, but rather to open up the very possibility for experiencing an environment at all.

This notion of the envelope, as something that both discloses and encloses, brings to mind Heidegger's notion of the open. Heidegger describes this metaphorically through the space of a clearing: an opening in a forest encircled by trees. He uses this metaphor of the clearing to understand how the world that appears to humans always operates in a dual process of concealment and disconcealment. As Krummel (2006: 415) describes it, an open 'is a clearing accompanied by darkness as delimiting contours, defining what are present (i.e. beings) via the withdrawal of the not present (non-being)'. Inhabiting an open means that the process of identifying entities always operates through an ongoing movement of differentiation between the present and the not present. However, an interface envelope does not create a clear distinction between presence and absence or an interior and exterior. Rather, an interface envelope emerges from the relationship between a set of objects which disclose qualities in order to create a space and time in which player activity takes place. As such, the resolution and technicity of objects work to envelop the player or user.

To understand the concept of envelopment, we can turn to McCormack's (2014) example of a hot-air balloon. Discussing how the silk of a hot-air balloon is used to hold a pocket of heated gas, he argues that envelopment is 'a technical process of experimental spherification in which new capacities to act are generated through folding materiality upon itself in a process of differentiation' (McCormack 2014: 7). He goes on:

> envelopment...[is]...a particular relation between different states of matter, a process in which matter folds in on itself, generating the potential

for variation in the power and properties of things. And, as Serres also reminds us, with the crafting of the skin of the enveloped thing, a turbulent field of atmospheric gas is 'organized into an exchanger'. And an exchanger allows the diffuse, the atmospheric, to pass into, and give volume to, the circumstantial qualities of the body of the local as a sensing, feeling actuality. (McCormack 2014: 8)

McCormack links this notion of folding to Serres' account of topology. Here, 'thinking topologically like Serres means understanding envelopment not so much as the generation of an object, but as the shaping of generative relation between what Serres would call two "phases of matter"' (McCormack 2014: 8). Law (2002: 95) defines topology as 'a mathematics that explores the possibilities and properties of different forms of continuous transformation'. To explain this complex area of theory, he suggests that topology is the study of shapes and their continuity: 'in topology...a shape is said to hold its form while it is being squeezed, bent, or stretched out – but only so long as it is not also broken or torn' (Law 2002: 94). To fold or envelop then is to recognize that 'the shape and size of things or the distance between them is less significant than what holds them together; that is, the ways in which they are connected, the nature of their relatedness, so to speak' (Allen 2011: 285). What is useful about this notion of topology is that it emphasizes that objects are deformable and malleable, while still retaining a capacity for opening, surrounding and enclosing. Topology calls this ability to bend, stretch and contort without breaking homeomorphism (Paasi 2011).

Taking aspects of these three different accounts of envelope and envelopment together, an interface envelope can be defined as a localized opening of space time, or emergent effect of the continuous transductions between a player's body and the technicity and resolution of objects they engage with when they use an interface system. Interface envelopes are not composed of particular forms of solid, liquid or gas, as in the examples above, but of homeomorphic spaces and times. Interface envelopes are homeomorphic in the sense that they actively bend, modulate and shift within a series of limits. This allows them to accommodate and respond to what players are doing while engaging with the interface. In doing so, the interface envelope can attempt to organize the habitual relations between the player's body and brain and their temporal perception.

Developing the concept of interface envelopes as localized foldings of space-time, the following section demonstrates how interface envelopes are designed to modulate the player's capacity for retention and protention through the specific example of *Final Fantasy XIII*. The third section suggests that this capacity for homeomorphic modulation be termed envelope power.

In turn, this section differentiates envelope power from the accounts of psychopower and neuropower on which I have been drawing so far. The fourth section seeks to show how a logic of the interface envelope has developed and become more sophisticated over time, by examining changes in three major game series: *Final Fantasy*, *Resident Evil* and *Metal Gear*.

## Homeomorphic modulation

Depraz (2004: 14) explains that, '"to modulate" means "to vary", "to be inflected", "to adapt" to particular cases or contexts of meaning'. According to Clough (2013: 116), 'it might be better to describe mediation as modulation, intensifying or de-intensifying rhythmicities and forces that are of, but also below, above and other than human perception'. In the context of envelopes, modulation means two things. First, it refers to the process of assembling and placing objects in an interface system in order to shape their resolution and technicity. Second, modulation refers to shaping players' habitual capacities to act as well as their perceptual capacities for anticipation and recollection. How designers attempt to modulate the resolution and technicity of objects in an interface environment and the temporal states and habitual practices these modulations generate is then intimately linked to the types of envelope that potentially emerge when a player engages with a game.

The process of modulating the technicity and resolution of objects in an environment to produce an envelope can be examined through a number of cases. Take, for example, the battle systems of the very popular *Final Fantasy* role-playing game (RPG) series and, in particular, *Final Fantasy XIII*. Like many Japanese RPGs, gameplay in *Final Fantasy XIII* is split between exploration, character interaction and battle. The battle portion of these games tends to take place in a separate system from the main game. When encountering an enemy in *Final Fantasy XIII*, the screen will transition from the main exploration environment to a different battle field where the player will fight and (try to) defeat the enemy. If successful, the characters will perform a victory pose. A status screen will appear detailing the experience and outlining the items received from the battle before the player is returned to the main exploration environment of the game.

The *Final Fantasy XIII* battle system operates around a modified active time battle (ATB) system. The game's designer, Yuji Abe, terms this 'the paradigm system'. In previous games, the player could individually control each of the characters' inputs during battle. The paradigm system involves assigning a specific role for each character to play in battle, which forms a paradigm. Each character has a number of roles, such as commando (whereby the character will focus on building chains of attack) or sentinel (where the character will shield allies from attack). While the player directly controls the main character called Lightning, the computer AI controls the other characters in the battle and decides what moves to use depending on the paradigm the player has employed. For example, the healing paradigm would involve two of the characters focusing on healing spells and the other casting defensive spells. Once fully healed, the player could then switch to a more offensive paradigm such as an attack paradigm, which would mean that the computer AI would then instruct each non-player character to employ physical or magical attacks on the enemy. In summary, rather than micromanage every move each character performs, as in the ATB system, in the paradigm system, the player acts to macromanage the battle by watching how the computer-controlled characters are performing against the enemy and then switching the paradigm depending on the situation. Abe describes the paradigm system in the following way:

> In the battle system of *Final Fantasy XIII* there are a lot of characters moving on screen at once, so we really wanted to make sure the battle system was something intuitive that the players could grasp right away so the experience wouldn't be lost. If players had control over [their] party members it could be a little overwhelming so keeping the experience with one main character was really intentional on our part.... (PlaystationBlog 2010: n.p.)

What is interesting about Abe's rationale for the paradigm system is that it is a solution to a problem that he himself created in the very design of the battle system. The frantic real-time movement of characters and flashy animation effects that accompany attacks and magical spells were deliberately placed into the game in order to keep the player stimulated in relation to what is happening. In Abe's words, 'the basic concept for this system is that it is speedy and tactical' and '[w]ith this battle system we wanted to give the characters lots of lively movement and make the battles very action orientated, but still keep the strategic element strong' (PlaystationBlog 2010).

Indeed, Abe did not just want to encourage generally speedy responses from players in a vague sense; he had particular ideas about the temporal windows in which action should be undertaken by the player. In the same interview, he argues:

> the most effective way to use the paradigm system is to...change it every 10–15 seconds or so...This is a good pace to switch the paradigms and

if you keep this pace it becomes easier for the ATB gauge to fill up, so characters can go into certain actions right when they switch because the ATB gauge will be ready for the shift. (GameSpot 2010: n.p.)

By default, the ATB gauge is split into three equally sized sections (although players can gain three more sections as the game progresses). The ATB gauge is a red progress bar that moves from left to right and automatically fills and refills after the player inputs commands during battle. Actions inputted by the player are presented in the form of blue rectangular boxes, which sit parallel to the ATB gauge on the screen. In order for the character the player controls to perform the specified action, there needs to be enough bar to cover the entirety of the rectangular box. As a result, the technicity of the now is shaped around the limits and movements of the ATB gauge and the speed at which it fills.

Here, the technicity of the ATB gauge works in both a quantitative and qualitative sense to shape the emergence of the now for the player. The bar moves seemingly smoothly and indivisibly across the screen, with the full six gauges filling in less than five seconds. At the same time, this movement is framed and delimited by the six markers that cut and split the gauge into discrete sections. When players are engaging in battle, they have to wait for the gauge to fill. They can anticipate when it will fill but are unable to do anything while it fills. By forcing players to wait for five seconds to input an action, the ATB gauge operates as a tertiary retention that is designed to generate anticipation (protentions) in players. Players do not have to actively count or wait for the right moment to attack because the bar denotes when they can attack. In doing so, the now emerges through the automated speed at which the bar fills and the anticipation of the player who is waiting to input a command. In *Final Fantasy XIII*, the ATB gauge creates a now based around a five-to-six-second loop of action where the players anticipate a now, enter a command and then wait for five seconds to input the next command.

Other objects in the interface system are designed to appear with a particular resolution. For example, in *Final Fantasy XIII*, the total number of hit points or health that an enemy has is displayed to the player graphically (in the form of a life bar) rather than numerically. When the player hits an enemy, the enemy's life bar will shorten. The amount of damage that the player has inflicted on their opponent is simultaneously displayed numerically around the enemy's body. The visual depiction of the points of hit damage is designed to work on the level of affect, rather than providing information the player could actually use to aid their strategy in battle. As Rogers (2010) suggests, the number of hit points that enemies have in the

*Final Fantasy* games has radically increased since the series' first iteration. In his words, 'the Final Boss of *Final Fantasy* had 2,000 hit points: we should have known that Average Enemy Grunt in *Final Fantasy XIII* would have at least 100,000' (Rogers 2010: n.p.).

The increase in enemy hit points across the *Final Fantasy* series could be partially understood through Heidegger's (2012) concept of the gigantic, in which he considers the logic of modern technology to encourage humans to understand entities in the world primarily through quantitative forms of calculation. However, in another sense, these numbers exist primarily as objects with a capacity to produce resolutions, rather than as representations of mathematical value. The resolution of these numbers can be understood through their graphical design and the ways they emerge and dissipate during battle. Upon hitting an enemy with an attack, an overlapping number will appear on the enemy's body. The number begins at zero and then quickly climbs to denote the amount of damage inflicted to the enemy. Once the number has stopped, it remains static on screen for a moment and then floats upwards away from the enemy and dissipates at the same time. When fighting multiple enemies, or inflicting multiple points of damage, the screen can quickly fill with these numbers, which appear, fade and overlap one another, often in quick succession.

The numbers that appear on screen in the *Final Fantasy XIII* battle system give the player's attacks a high resolution by depicting how effective these attacks are and showing how they have affected the enemy that the player is attacking. By spacing these numbers out and placing them in relation to the enemy, the player's attacks transduce secondary qualities of weight and connection that might be absent within other RPG battle systems. The way large sets of numbers appear alongside one another is used to attempt to transduce sensory overload in the player's body. Unable to pay attention to the specific numbers that emerge on screen, the massiveness of the damage caused is transduced into a sense of being overwhelmed by the speed at which the numbers appear and the impossibility of adding them together to create some manageable coherence out of them. Creating a sense of weight and overload is desirable from the game designers' point of view because it generates an illusion that the player is part of the battle, when in actual fact, success or failure is based on numerical calculations and AI decisions that the player has no real control over.

By constructing and placing objects alongside one another and encouraging the player to engage with these objects, an envelopment can emerge. Thinking back to McCormack's account of the hot-air balloon, this envelopment serves as a partial enclosure in which space and time appear in a particular way in order to disclose certain kinds of action as more or less possible or desirable. McCormack (2014: 7) argues processes of envelopment should be considered topologically. Here, partial enclosure is understood as:

> stitched along the pleats of matter, along the seams and skins of technologies of envelopment through which emerge discrete, fabricated presences held in place by differences in density rather than structures or scaffolding. This is a topological process, in which 'hand and gaze devote themselves to connecting the far and the near, or to creating varieties from a simple line: flat or voluminous, tight or loose, dense or scattered'.

The envelope produced by *Final Fantasy XIII* is an envelopment in which space and time appear and are orientated around the immediacy of player action. This is intended to keep the player interested and concerned with what is happening within the battle. Objects such as the ATB gauge and damage information offer forms of sensory differentiation that make time and space appear as narrowed, focused and fast moving. As the player flicks their eyes between the ATB gauge and the damage numbers on screen, the relationship between here and there, then and now becomes folded into one another.

Drawing upon McCormack's topological language, space and time become tightened and loosened as the technicity and resolution of objects shift within the interface envelope. For instance, as the camera zooms in on the damage markers created by player attacks, the resolution of the attack increases, pulling space together. As the camera pulls back and pans automatically around the ongoing battle, space and time loosen a little, offering the player a sense of an environment beyond the immediacy of battle. At the same time, players are switching between paradigms in order to gain some tactical advantage over the enemy. At any one moment, a number of qualities are emerging from the transductions between objects, which results in the player being affected on a number of different levels and having to differentiate between a variety of objects and processes within different temporal windows.

Stiegler's concepts of perception, retention and protection allow us to understand how the envelopes of battle systems in games such as *Final Fantasy XIII* are actively designed to try to narrow the relationship between experience, memory and anticipation to create a continuously modulating present tense. In Stiegler's terms, the *Final Fantasy XIII* battle system attempts to amplify the user's capacity for both protection and retention by closely wedding together primary and secondary retentions (perception and memory) which are required to project (anticipate) future events. The technicity and resolution of objects that make up the battle system of *Final Fantasy XIII* attempt to link anticipation and recollection around particular metrical boundaries – the five–six seconds it takes to fill the ATB gauge, the one–two seconds for which attack data appears before disappearing and the ten–fifteen seconds between paradigm shifts. In doing so, the battle system encourages the player to concentrate on the present moment, but it is only able to do so because players must constantly shift between a state of anticipation (protention) and memory (secondary retention) that shapes how the moment of perception (primary retention) emerges.

Crucially, the production of this present tense is not a fixed form of presentness, but actively modulates around spatial and temporal limits that shift both within and across different games. In other words, interface envelopes are elastic rather than static and do not necessarily produce a uniform or generalizable form of present tense for players. This is because different videogames consist of a broad range of different inputs and outputs, which generate a broad range of qualities and, in turn, modulate a broad set of technicities and resolutions that shape perception within different envelopes of potential. At the same time, the emergence of envelopes is also influenced by the skill and experience of the player. While attacking a regular enemy in *Final Fantasy XIII* may be more or less automatic for an experienced player, who understands the paradigm system and knows that foe's particular elemental weakness, a difficult boss may require a more conscious strategy. Even within the same encounter, players may move between automatic habits, such as tapping the A button to queue up attacks, and more conscious moments where they try a new skill or spell for the first time.

All objects express both a technicity and resolution. The technicity and resolution of objects work to time and space the interface into a more or less coherent and navigable environment for the player. The modulation of this technicity and resolution is both a product of particular objects created by the designers (such as the stun meters, which encourage players to shift paradigms) and the contingent actions of players as they make sense of and engage with these environments. For example, *Final Fantasy XIII* players I have observed regularly tap the attack button repeatedly while waiting for the ATB gauge to fill. While the designers did not intend them to perform this action, tapping on the button also contributes to the production of an envelope by linking exterior somatic movements to the structures of anticipation designed into the game's systems. Tapping the attack button operates as a kind of release for the anticipation of what is about to happen and, in doing so, further folds and envelops the player's body into what is happening in the interface environment.

Designers also use indirect forms of encouragement to generate and further cultivate the production of envelopes outside of the particular systems from which these envelopes emerge. In *Final Fantasy XIII*, this takes the form of a results screen displayed after each battle. This results screen includes displays of a target time in which the designers expected you to complete the battle, the actual duration of the battle undertaken by the player, the points the player scored per second during the battle and a rating of the player's performance in the battle (out of five stars). The faster the battle is completed, the better star rating players receive. Hiroyuki Ito, a key engineer and inventor of both the turn-based and active-time battle systems in the *Final Fantasy* series, suggests receiving a bad score:

> may want...[the player]...to make more offensive paradigms to help...[them]...plough through battles in a timely manner...the merits of completing a battle with high scores would definitely be you earn more tactical points and also enemies will leave you with better items if you have higher stars at the end of the battle. But these are small merits compared to the satisfaction players get from achieving the type of scores that players envisage for themselves. The score is there to create a competitive atmosphere so that they can strive for something better and achieve that challenge. (Parish 2010: n.p.)

Designers build this kind of technique because they can't simply force or conjure envelopes into being; envelopes require the active and willing engagement of the player. As such, games designers use techniques such as the results screen in *Final Fantasy XIII* as an indirect form of modulation to attempt to encourage the generation of an envelope.

## Envelope power

When envelopes are successfully generated, they produce a form of power that can be termed envelope power. In Chapter 3, I linked the concept of resolution to neuropower through the way that objects are designed to construct habitual relations between body, brain and interface. Then, in Chapter 4, I linked the concept of technicity to psychopower through the way objects in the interface shape an experience of the now in order to capture attention. Successfully constructed interface envelopes generate an envelope power, which is distinct from either neuropower or psychopower (even if it is shaped by them). Envelope power differs from neuropower and psychopower in three main ways.

First, envelope power is more dynamic and amorphous than either neuropower or psychopower. Envelope power is homeomorphic, which is to say that envelopes are shifting localizations of space-time that emerge from the active relationship between bodies and objects that make up the interface environment. As I discussed in Chapter 4, Stiegler ultimately sees psychopower as about manipulating attention and memory by divorcing the present from a broader sequence of past and future now points. In contrast, envelope power works to modulate the limits around which the past and future appear as modes of potential. The present now is not cut off from some abstract past or future nows but is productively linked to particular pasts or futures that the designers want to push the player towards or away from.

Second, envelope power does not attempt to create homogenized populations of individuals who all possess very similar somatic capacities and memories as one another. This contention is more or less explicit in both Neidich's and Stiegler's accounts of neuropower and psychopower. Stiegler (2010a, 2013a, 2013b) is critical of how psychopower utilizes media to standardize and quantify the processes of memory selection. For Stiegler, beings are constantly in process; they are assemblages spread across and individuating between a variety of materials and components. To be an individual then is to both retain aspects of the world and to be retained by aspects of the world. Stiegler considers the point of selection to be key to the process that differentiates one individuated being from another. In Stiegler's (2011a: 112) words, 'individuation always consists in a selection.... That which came to pass is that which came to me, and what I retain is therefore that which constitutes the singularity of my experience: what I retain is not what my neighbour retains'.

For Stiegler, the current media situation is characterized by an ongoing tension between individuation as always singular on the one hand and an attempt by the culture industries to homogenize the primary (and thus secondary) retentions that constitute individuation on the other. As Stiegler (2011a: 112) puts it, 'I only think differently from others, I only feel differently from others, I only desire differently from others.... In short I only exist – because of the retentional process in which I consist is unique'. Stiegler (2011a: 112) goes on to suggest that 'even when we live through the same event... we have different experiences: we have previously accumulated differing primary retentions that have meanwhile become our pasts, that is arrangements of secondary retentions'. While two people can experience exactly the same media event and still take a different singular experience from it, the culture industry aims 'to reduce the differences between primary selections, that is, to intimately control the process by homogenizing individual pasts' (Stiegler 2011a: 114). In other words, while each retention of an event is singular, these are all singularities based upon the same mediated event, such as a television programme, which means multiple individuals' secondary retentions become similar to one another. This creates a situation in which 'behaviours are ... more and more woven by the same secondary retentions and tend to select the same primary retentions, and ... [so] ... increasingly resemble one another' (Stiegler 2011a: 118).

This is problematic because these secondary retentions are largely created and controlled by the culture industries for the sake of profit. This creates homogenized forms of individuation and, therefore, homogenized individuals who are unable to think critically for themselves. Stiegler is very clear about this in several places. He claims that 'the function of the culture and programming industries is to take control of ... processes constituting collective secondary retentions. This control is achieved by replacing inherited pre-individual funds with what the culture and the programming industries produce, and through this substitution to cause the adoption of retentional funds conceived according to the needs of marketing' (Stiegler 2011a: 113).

In a similar manner, Neidich suggests that neuropower works to create populations of bodies with the same neural architecture, who will respond to marketing and advertising in predictable, controlled ways. As he puts it, neuropower 'administers the pluripotential of neuroplasticity in the curating of a homogenous people both in the present and future' (Neidich 2010: 539).

Instead of attempting to homogenize or minimize the difference between the somatic capacities and memories of different bodies, envelope power actively attempts to cultivate bodies with differentiated and distinct capacities. Creating bodies with differentiated capacities and memories is beneficial to the new media and videogames industries because monetizing different memories and capacities forms the business model of many new media services. For example, social media requires people to post about their own experiences or upload photographs; similarly, online videogames require players with different skill levels in order to create the possibility for competition.

Third, envelope power does not simply diminish capacities for attention and the ensuing forms of cognition this attention allows. Instead, envelope power actively cultivates the generation of new capacities to act through processes of homeomorphic modulation. For Stiegler (2010a: 55), psychopower is about creating a mode of stupefied attention, which he suggests leads to an outright 'destruction of consciousness'. In his words, 'the crippling limit of ... attention control apparatus, is that it destroys attention itself, along with the ability to concentrate on an object of attention' (2010a: 13). However, envelope power is not about breaking down a capacity to concentrate on an object. Instead, it is about increasing the players' capacities to concentrate on an object. In videogames, attention is not manipulated in order to sell available brain power to advertisers (as it is in Stiegler's narrative). Rather, envelope power works to increase players' capacities to sense difference between units of space and time that are increasingly micro-differentiated. By encouraging and cultivating this capacity for micro-differentiation, envelope power opens new synaptic territories from which the videogame industry can derive economic value. In this regard, envelope power is closer to Neidich's account of neuropower than Stiegler's account of psychopower. Neidich distinguishes between noopolitics and neuropower by emphasizing the active nature of neuropower. As Neidich (2010: 543) puts it, 'noopolitics relationship to sensibility is defined as a passive tele-visual like process and concerns seduced attention leading to passive tele-visual memory. Neuropower has shifted this engagement of the subject to one that is now active and mobile'.

But, while he recognizes the active nature of neuropower, Neidich still suggests that neuropower is ultimately incapacitating. In relation to Global Positioning System (GPS) devices found in phones, for example, he argues that GPS leads to a 'capturing and externalization of thought at the expense of deep internalized contemplation' (Neidich 2014: 281). For Neidich, the tertiary retentions of GPS 'produces a dangerous situation for the body on two fronts. First, the corporeal body in its disengagement is left defenseless and subject to motor vehicle accidents. Second, the mind, which in spite of the obvious positive side effects of GPS devices of spatial knowledge and location awareness is subject to the effects of disuse' (Neidich 2014: 282). Different from neuropower or psychopower, envelope power is productive in two senses. It requires and encourages the active engagement of bodies, and it also cultivates new capacities to act. As such, envelope power is not repressive but empowering, which makes it all the more malleable and difficult to undo or reflect upon.

As such, the types of envelope and envelope power that videogames can generate should not be confused with the effects of interfaces such as gambling machines that can produce numbed or addicted bodies. In her analysis of video-machine gambling, Schüll identifies a zone of perception that problem gamblers attempt to enter, which she calls the machine zone. She defines this zone as 'a zone in which time, space and social identity are suspended in the mechanical rhythm of a repeating process' (Schüll 2012: 13). In this zone, players lose all sense of self as they are drawn into a perpetual present of betting, until their resources are extinguished.

Schüll goes as far as suggesting that the perpetual present created by the machine zone can directly lead to addictive tendencies in which machine play becomes compulsive. As Schüll (2012: 132) argues, 'addiction emerges from an ongoing process of human-machine adaptation in which...changing needs (i.e. what...[is]...required to enter the zone) are spurred by the continual "matching" of those needs by design'. This machine zone is enabled by a perfect contingency whereby a player becomes focused on seeking to anticipate what will happen next and learns to respond appropriately to this anticipated event. Schüll (2012: 171) defines perfect contingency as a point at which the player's actions 'become indistinguishable from the functioning of the machine ... a kind of coincidence between ... [the players] ... intentions and the machines responses'. Perfect contingency generates a 'perpetual present tense of the play zone, a zone characterised by affective adaptation rather than analytic leverage' (Schüll 2012: 155).

While Schüll's account of the machine zone might seem similar to the theory of the envelope I have been developing in this chapter, it is actually quite different. The machine zone Schüll (2012: 52) describes is narrowly based on 'accelerating ... extending ... and increasing' the players rate of play through 'compress[ing] ... the greatest number of physical gestures into the smallest unit of time' (Schüll 2012: 57). In contrast, interface envelopes operate by modulating the relationship between gesture and perception around shifting units of space and time to extend the length of players' engagement with a game, while maintaining their attention and interest. Rather than trying to simply speed player's input up, games such as *Call of Duty*, *Street Fighter* and *Final Fantasy XIII* require players to develop exacting modes of engagement and often involve complex forms of contingency that are irreducible to a simple quantitative calculation. In other words, interface envelopes require skill and a form of creative anticipation that the machine zone of interfaces such as gambling machines discourage.

The kinds of habit that are generated and contracted in the machine zone that Schüll describes are also very different to the perceptions and habits contracted in an interface envelope. As Schüll (2012: 135) states, the machine zone 'is attainable only at the threshold where rhythm holds sway over risk, comfort over perturbation, habituation over surprise'. Whereas the machine zone encourages players to enter into a space where action is performed in as unreflective and automatic a way as possible, envelope power encourages a more complex and continual movement between reflective and automatic response. Furthermore, Schüll (2012: 13) considers the zone as a perceptual state 'in which time, space and social identity are suspended in the mechanical rhythm of a repeating process'. Envelopes are not just a perceptual mental or bodily state in which space and time are suspended, but are openings and envelopments of actual space and time through which a particular modality of perception is cultivated.

Comparing the concept of machine zones to the interface envelope using Stiegler's language of retention and protection, we might argue that the machine zone attempts to discourage and diminish capacities for both retention and protection. In videopoker, playing as quickly as possible encourages players to ignore previous plays (secondary retentions) in order to focus on a now that is determined by the speed of the mechanisms and electronics of the machine to display the next set of cards on screen. The next set of plays is only a moment away, which means the player's anticipation (protention) is focused on a future now, which may occur in only one second's time. Videogames modulate the technicity and resolution of objects in the interface environment in order to create a much more dynamic and shifting relation between retention and protention. Rather than attempting to fix retention and protention around a static set of spatialized markers and inputs (such as the fixed rows of cards on the videopoker machine and the single button used to operate the machine), game designers generate gameplay mechanics that shift and move these markers around. In *Final Fantasy XIII*, hit points are not displayed in a static location of screen, but emerge and dissipate in overlapping fields around and in front of the enemies the player's avatar is attacking. While the form of 'presentness' in the machine zone of gambling machines is very narrow and uniform across the play session, videogame envelopes are far more dynamic, involving complex forms of creative anticipation and somatic skills.

Interface envelopes and envelope power are not about creating addicted, passive or unthinking bodies. Instead, envelope power can be defined productively in two ways. First, envelope power works to increase users' habitual and embodied capacities to sense difference and modulates these capacities to encourage both automatic and reflexive modes of response. Second, envelope power draws upon players' habitual capacities to sense difference to alter the relationship between their temporal anticipation and recollection in order to encourage continued engagement and consumption of the products they are using.

## Shifting logics of the envelope in games design

The turn towards creating envelopes that attempt to modulate the relationship between players' capacity for perception, retention and protention is part of the logic of the interface envelope that is becoming more and more apparent within particular aspects of big budget and AAA videogame design. This change has been taking place over the previous three decades or so, but has intensified in the last ten years. There seems to be at least two interconnected reasons for this intensification. The first is that videogame's companies are increasingly attempting to formalize a 'theory of fun' (Koster 2005) and understand how to reliably create game mechanics and environments that players will find interesting. This can be evidenced through the rise of a whole sub-industry of testing where techniques and theories from psychology and neuroscience are being utilized to study how players engage with and react to games. The results of these tests are directly inputted back into the design process (Heijden 2010). Second is the rising cost of videogame development (Chapple 2014). A game for the original Sony Playstation (released in 1994) might cost between $80,000 and $1.7 million to develop, while a game for Playstation 2 (released in 2000) cost between $5 and $10 million (Leach Waters 2005). In 2014, a Playstation 4 game can now cost anywhere between $50 and $500 million to develop and market (Grover and Nayak 2014). This exponential rise in development costs means that the industry is increasingly risk averse. With such high stakes in mind, games publishers and companies minimize the risk by making sure the game mechanics and environments designers create capture and hold players' attention as effectively as possible in order to recoup development costs and generate profit.

Understanding how the logic of the interface envelope has become incorporated into games design can be evidenced by examining changes between games in a number of long-running series. These series are *Final Fantasy*, *Resident Evil* and *Metal Gear*. The purpose of each of these examples is to show how different series can create envelopes using objects with different resolutions and technicities, while still trying to create envelope power. These examples also show both the diversity of envelopes that can be generated and emphasize that envelopes and envelope power are not limited to specific genres of game or videogame consoles.

The *Final Fantasy* battle system has had many iterations and versions since the original *Final Fantasy* game was released for the Nintendo Entertainment System in 1987. The original *Final Fantasy* game had what has come to be known as a turn-based battle system. In this system, the player commands a team of allies, who occupy one side of the screen, in a battle against enemies, who occupy the other side. The player-controlled characters and the enemies (who are controlled by the game's artificial intelligence or AI) then take turns to attack one another. The player uses a pointer to choose from a series of menu-based options, such as 'fight', 'magic', 'item', etc. They, then, select an enemy or ally on which to carry out this command. As players destroy more enemies, they gain more experience, which in turn unlocks new abilities or magic spells.

Over the series of fifteen or so mainline iterations, this basic turn-based system has been modified a number of times. While the first three *Final Fantasy* games used a traditional (unnamed battle system), *Final Fantasy IV* to *IX* utilized an ATB system. *Final Fantasy X* used a Conditional Turn-Based Battle system, *Final Fantasy XI* used a Real Time Battle system and *Final Fantasy XII* used an Active Dimension Battle system. As the names of these battle systems imply, each iteration introduced more real-time and active elements, which require players to respond in context-dependent ways to what the enemies are doing as they are doing it. *Final Fantasy IV* introduced a graphical bar in the menu area of the battle screen that filled in real time. When their bar is full a character can attack; the player does not have to wait until it is their turn before attacking (as they would have in previous games in the series). In *Final Fantasy XII*, the system of turn-based battles was completely done away with (although the basis of the turn-based system returned in *Final Fantasy XIII*), and players approached and moved around their enemies as well as giving commands to their characters in real time, without the need to transition to a separate battle mode.

Hiroyuki Ito suggests that a turn towards real-time battling was about creating the illusion that the player had more control and influence over the battle system than they actually did:

> I felt that if we put too much of an action element in the game, that would alienate users. The theme I came up with was an action-like game with no action elements. [laughs] ... Something without the hardcore reflex action elements.... But even if it's automatic, you have to give people something to do, or else they'll just get bored. So I made the processes automatic, but I tried to give players something to do – like stepping on the accelerator or stepping on the brake – so it felt like they were part of the process.... At the time, there was a shift going on in Formula One racing, where they introduced semi-automatic transmissions. People are shifting gears, but there's no clutch pedal in the car. It's half of the process of the manual transmission built into an automatic. I felt that by drawing inspiration from that kind of system, implementing that kind of system would give people the illusion that they're doing much more to drive the action than might actually be real, because of a lot of the processes are automatic. (Parish 2010: n.p.)

As is clear from the above quote, the battle system in the *Final Fantasy* series has evolved to modulate players' habitual capacities to act as well as their temporal perception. While many of the underlying processes in the battle system are automatic, the development of the battle systems across various iterations of the *Final Fantasy* series has attempted to modulate habitual capacities and temporal perception in increasingly dynamic and context-sensitive ways. By ramping up the numbers of variables the player has to track – the amount of information presented on screen alongside the number of hit points attributed to each enemy – there is a clear desire by the designers to try and create an interface envelope that draws players' attention into a narrowly defined, albeit continuously modulating, present tense in ways that were simply not possible in the original *Final Fantasy*.

Similar evolutions in interface design and the resulting envelopes they produce can also be charted in other long-running series, such as *Resident Evil*. One can clearly see the difference in the designs of *Resident Evil* compared to *Resident Evil 4* and how these differences alter the types of envelope that potentially emerge. One key difference between *Resident Evil* and *Resident Evil 4* is the resolution of the weapons available to the player in each game. Weapons in *Resident Evil 4* generally have a much higher resolution than the weapons in *Resident Evil*. This is due to a number of factors. The first is that weapons in *Resident Evil 4* are modelled using a higher number of polygons than in *Resident Evil*, which provides the player with more visual detail. The second is that weapons in *Resident Evil 4* are upgradable, meaning that they have a high degree of homeostasis because they are made up of multiple parts, and changing these parts can affect the capacity of each weapon as a whole.

Alongside the changes in the resolution of weapons, the number of enemies within an environment in *Resident Evil 4* is also largely increased compared to the original game. In key sections of *Resident Evil*, players might be shocked to see three zombies in the same area. However, in *Resident Evil 4*, the player regularly encounters six to eight zombies within one environment. Furthermore, the capacities of these enemies also changed markedly between *Resident Evil* and *Resident Evil 4*. Whereas the unarmed zombies in *Resident Evil* stumble slowly towards the player's avatar, the higher resolution zombies in *Resident Evil 4* communicate through speech, carry and use weapons, such as pick axes or sticks of dynamite, and are capable of running and climbing. These changes markedly altered the encounter rate and the number of enemies that the player has to try and deal with at any one moment.

The resolution of the weapons, alongside this increased number of enemies, alters the technicity through which the now appears. The low resolution of the avatar and weapons, alongside the low encounter rate of enemies in *Resident Evil*, creates a now that modulates around a widely spaced set of retentions and protentions. Without the need to focus on a high number of enemies, gameplay appears slower and more ponderous to the player as their attention is not continually forced to shift between objects in the interface environment within tightly delimited windows. In *Resident Evil 4*, this envelope is drawn around a much more narrowly differentiated set of retentions and protentions. When encountering the multiple enemies that make up the majority of scenarios in *Resident Evil 4*, the player's perception is organized around not only killing that particular enemy, but also selecting which enemy to tackle next based upon their own assessment of which enemy poses the most threat.

Anticipation is also modulated by the memory (secondary retention) of how many shots it takes to kill the enemy and how long reloading takes (tertiary retention). When shooting enemies in *Resident Evil 4*, players can stay focused on their current activity without having to remember how many bullets they have left in a weapon because, unlike *Resident Evil*, there is an ammunition counter on the Heads Up Display. Externalizing this information in the form of a counter on screen is another way of spacing secondary retentions and protention to generate a narrow now point. It would be possible to argue that all the games in the series attempt to generate interface envelopes and, therefore, envelope power. Yet, it is also clear that, as the series has developed, the mechanisms used to generate envelopes have become more sophisticated, so as to produce envelopes with more dynamic limits and with a broader capacity for homeomorphism. In other words, the objects in *Resident Evil 4* allow designers to attempt to loosen and tighten the envelope both more narrowly and broadly than in *Resident Evil*, which allows the more effective production of envelope power.

The *Metal Gear* series serves as another clear example of how the envelopes that games attempt to generate have changed over time. The *Metal Gear* games are military action games that emphasize stealth. Within all of the *Metal Gear* games, the player is given very few resources and so has to attempt to hide from and evade the enemies they encounter. While direct combat is possible, it is generally discouraged outside of key set pieces or boss battles. The original *Metal Gear* game was released in 1987 for the MSX2 computer platform, with the sequel, *Metal Gear 2: Solid Snake* released in 1990 for the MSX2.[1] Later games in the series included *Metal Gear Solid* (1998), *Metal Gear Solid 2* (2001, Playstation 2), *3* (2004, Playstation 2), *4* (2008, Playstation 3) and *V* (2014, Playstation 4). Examining these games in detail, we can identify how they were designed in an attempt to produce different kinds of envelope.

While the original *Metal Gear* and *Metal Gear 2: Solid Snake* shared the same top-down isometric environments, a number of objects in the interface differentiated the kind of envelopes these games potentially produced. In *Metal Gear*, enemy guards appear in a low resolution and can only notice the player if they are directly in the enemy's line of sight. This created a situation

---

[1] This analysis ignores the NES versions of the first two games, Metal Gear (1987) and Snakes Rovongo (1000). While also developed by Konami, neither of these games was designed or endorsed by the series creator Hideo Kojima, and they are generally considered inferior to the MSX2 games.

in which the player's avatar can be touching against the enemy without the enemy noticing. This means players can be brazen in their movements and run around enemies with little fear. In *Metal Gear 2: Solid Snake*, the resolution of the enemies was increased by giving them a 45-degree cone of vision. If the player manoeuvers their avatar, Snake, into this cone, the enemy becomes alert and attacks. In addition, enemies in *Metal Gear 2* can move beyond the confines of a single square of the map (delimited by the edges of the visible area of the screen). To support this, the player is provided with a Soliton Radar. The radar displays the navigable space of the map (presented in black) and objects (presented in green). The player's position is represented as a red dot and the enemies are represented as a white dot.

Alongside the addition of the Soliton Radar, *Metal Gear 2* also introduced a specific alert phase. When enemies notice the player, the radar jams and the player has to hide. If the player successfully evades the enemies during the alert phase, the game returns to its normal state and the Soliton Radar is reactivated. The envelope in *Metal Gear 2* then focuses protection on the continuing fear of being encountered, which is shaped by the resolution of the enemies in the environment. The game also creates tertiary memory, through the radar, which externalizes enemies' position once they leave the immediate screen. This means that players do not have to mentally keep track of their position and can focus on avoiding one enemy at any particular moment. Here, protection revolves around a series of entities, such as the enemy troops which are moving around the environment in which the player's avatar is immediately located as well as multiple other screens, represented on the Soliton Radar. In practice, this can result in the players' attention being continually pulled and pushed from the Soliton Radar to the environment and back again, as they attempt to anticipate when enemies will emerge on screen in order to avoid entering the alert phase.

Modulating the relationship between tertiary retention and protection in *Metal Gear 2* creates a fairly narrow now point in which action is orientated around feedback loops that occur and resolve themselves within ten to fifteen seconds. This is quite different from the original *Metal Gear*, where the objects in the interface work to create quite a broad envelope. The player can work out very quickly that enemies remain within their square of the map and have a very limited line of sight, rather than a broad cone of vision. There is also no alert phase. This means the player does not have to pay attention to enemies as closely or anticipate the emergence of potential alert modes. This creates a less-pressured situation in which the player has a more active capacity to reflect on circumstances outside of the immediate present, meaning the relationship between protection and retention is potentially weakened and widened.

*Metal Gear Solid 3* did away with the Soliton Radar system as a standard piece of equipment and left the player to try and work out where enemies were using environmental cues, such as enemy speech, footprints on the ground and entering into the games first-person view mode to scout areas using a pair of binoculars. This shift away from an overall perspective on the map, in which the player clearly knew the presence, location and field of view of enemies, to a much more partial and fragmented viewpoint altered the kind of envelope the game potentially generates compared to earlier games in the series. Removing the Soliton Radar in *Metal Gear Solid 3* further altered the relationship between protentions and retentions as the player now has to constantly anticipate an enemy who is not clearly located or represented for the player on a map. This anticipation alters the structure of players' actions, drawing them into a narrow envelope in which action is organized around small amounts of movement. Rather than being able to run around the map using the Soliton Radar to avoid each enemy's cone of vision, players have to move slowly, stop and check the environment for potential enemies before they can move again.

*Metal Gear Solid 4* further developed this dynamic. Neither does it externalize and spatialize enemies' location as a tertiary retention within a radar interface separate from the main environment, nor does it remove such tertiary retentions altogether. In *Metal Gear Solid 4*, a semi-transparent ring surrounds Snake when he is crouched or lying on the ground. The ring is usually flat against the ground, but peaks in the direction of an enemy or other threat. The ring helps to create a further narrowed interface envelope by allowing players to focus their concentration at one point on the screen. When the ring spikes, this encourages players to attempt to locate and pinpoint the source of the threat. The player has to sneak around, shifting the camera perspective, rather than simply stand still and use the radar to see the enemy's position represented as a dot on the radar screen (as in *Metal Gear 2: Solid Snake* or in *Metal Gear Solid 1* or *Metal Gear Solid 2*). The threat ring also serves to increase the overall pace of the game. When the ring is flat, players know they can safely move through and explore the environment without fear of attack by the enemy. This helps avoid players becoming frustrated because they feel they have to sneak carefully everywhere only to discover that there were no enemies in the area in the first place. The introduction of the threat ring in *Metal Gear Solid 4* attempted to mitigate some of the frustrating effects of the very narrow envelope of cautious action encouraged by *Metal Gear Solid 3*. By showing the player that an area was clear (the threat ring is flat), the designers modulated the speed at which players potentially moved, thus altering their relationship between protention and retention through which a now emerges. Rather than continuously moving slowly, players could explore at ease in particular sections, but then return to their more cautious modes of movement when they were alerted by the threat ring.

The most recent iteration of the series, *Metal Gear Solid V*, does away with both the Soliton Radar and threat ring of the earlier games. Instead, it employs a marking system, in which the player can use a set of in-game binoculars to focus on an enemy. After two seconds of focusing the binoculars in this way, a red triangle appears above the enemy's head alongside a distance marker. Once marked, enemies appear as a blue x-ray-like outline when they move behind obstacles, making it much easier for players to track their position and heading. Here, the resolution of objects in the environment is very high, with many objects disclosing information to the player about possible threats directly. The active, tertiary retention of real-time tracking in *Metal Gear Solid V* enables a homeomorphic envelope, which is more dynamic than either *Metal Gear Solid 3* or *Metal Gear Solid 4*. This is because *Metal Gear Solid V* encourages players to cultivate and continuously shift between two modes of engagement with the environment. When players first enter an environment, they are encouraged, upon fear of discovery, to carefully scan the environment to mark enemy troops. Once these enemies are targeted, the player can be less cautious as the marking system provides tertiary retentions that help players avoid running into an enemy and entering an alert phase. In doing so, the relationship between memory and anticipation is continuously opened and closed as players enter new areas, tightening as players sneak through the environment to target enemies and then loosening once all the enemies are targeted and the player can move more freely.

Although *Final Fantasy*, *Resident Evil* and *Metal Gear* are very different genres of games, created by different companies and designers, what connects their changes in design philosophy is the way in which they attempt to generate envelopes with increasingly high levels of homeomorphic dynamism organized around smaller and more micro-differentiated now points. Creating these micro-differentiated now points is key to generating the envelope power that makes them successful products. Of course, *Final Fantasy*, *Resident Evil* and *Metal Gear* (while popular) form only a small number of the games released over the last thirty years. With this in mind, I am not arguing that a shift towards trying to actively modulate players' capacities for retentions and protentions around a modulating now point is apparent within every game released. Far from it. However, I am suggesting that this is an important tendency that has emerged in response to the increasing risks involved in developing big-budget games, where the notions of fun and engagement are being formalized around a capacity to capture and hold the players' attention.

## The contingency of envelopes

While the examples of the *Final Fantasy*, *Resident Evil* and *Metal Gear* series are meant to demonstrate how logics of the envelope have changed over time, I am not arguing that these games always *do* produce specific interface envelopes that orientate players' experience of time and space around a fluctuating now point. Rather, they show a particular tendency in games design in which a narrowly structured interface envelope is the desired outcome for the games producers, even if players ultimately engage with games in ways that form alternative envelopes from the ones that the designers intend. It is important to recognize that interface envelopes are not assured outcomes of the relationship between body and interface, but a fragile and contingent phenomena that are always on the verge of collapsing or falling apart, dependent on the players' actions with the interface environment. As McCormack (2014: 16) argues, envelopes are best understood as a 'process of partial enclosure always open to the force of exchange'. Attempting to create an envelope is the designers' solution to the problem of shaping the players' perception in light of the contingency of their actions. But, it should be remembered that these actions often exceed the control of the designer. The contingency of these envelopes is evidenced by the difficulty that designers have in creating the conditions for their emergence (Ash 2010). In turn, this is reflected in the many games that fail, either commercially or critically, or on both accounts, precisely because players are not enveloped by the objects that make up the interface environment.

Developing an account of interface envelopes and envelope power allows me to modify accounts of psychopower and neuropower. Rather than a matter of attraction or distraction, the attention or neuro-economy can be rethought as an enveloping of life through the interface to modulate space and time to shape players' perceptual and habitual capacities. An account of envelopes problematizes existing accounts given by writers such as Rushkoff (2013) who argue that digital interfaces and new media lead to a distracted present, divorced from a broader context of pasts and futures. An envelope economy is about an increased focus on the present, brought about by an increased capacity to sense difference through a training of the habitual sensibilities of users of these interfaces.

Interface envelopes do not simply produce a perpetual present tense, but actively modulate the relationship between secondary retention (subjective memory) and protention (anticipation), through which a narrowly differentiated now point emerges. These envelopes emerge by creating tertiary retentions that short cut or stand in for secondary retention and provide constant stimulus that focuses the users' actions around a contingent, yet controlled, future. Through the creation of these envelopes, the now becomes increasingly differentiated and cut into smaller and smaller segments, with the primary goal of creating and squeezing economic value from the somatic capacities of the human body. Envelope power works to both train users how to perceive the world, and to create and generate value from this perceptual training and honing. But, envelope power does not end there. Game designers also use a variety of affective techniques to shape the emotional sensibilities of players outside of a particular engagement with game envelopes. While interface envelopes are affectively inflected, these ancillary affective architectures use statistical tracking systems on smartphones and tablets to encourage the player to feel connected to the game world even when they are not directly playing the game. Understanding how these affective architectures operate is the subject of the next chapter.

# 6

# Ecotechnics

*Measuring cannot but modulate and change the intensity of potential or affect. As such, the metric of measure necessarily will change with each and every measure. In this sense, the measure is an aesthetic measure or affective measure, understanding aesthetic measure to be singular, nongeneralizable, and particular to each event, or each modulation of potentiality.*

CLOUGH (2013: 119)

Discussing the role of digital labour, Clough argues that it is not helpful to reduce measurement to a process of quantification. She suggests that the primary role of measurement in cognitive capitalism is not categorizing and quantifying previously qualitative phenomena, but creating new affects through the very act of measurement. Clough's claims feed into the growing body of work that recognizes that affect is a key medium through which to create economic value in products. Extracting value from affect is not about encouraging people to buy into particular beliefs about the world, but about fundamentally restructuring how they sense and shaping how they feel through the ways in which they are affected by an interface. In Shaviro's words (2010: 3), digital interfaces can be understood as 'machines for generating affect, and for capitalizing upon and extracting value from, this affect. As such they are not ideological structures, as an older sort of Marxist criticism would have it. Rather, they lie at the very heart of social production, circulation and distribution'.

Much of the literature on affect and capitalism draws upon concepts of affective labour. The term 'affective labour' is developed from the work of Hardt and Negri (2000: 282–283), who define the term as a subset of immaterial labour:

> immaterial labor is the affective labor of human contact and interaction. Health services, for example, rely centrally on caring and affective labor,

and the entertainment industry is likewise focused on the creation and manipulation of affect. This labor is immaterial, even if it is corporeal and affective, in the sense that its products are intangible, a feeling of ease, well being, satisfaction or passion.

Developing this understanding of affective labour, Clough et al. (2007: 67) suggest that affect is not simply 'in' the body of the labourer; it is a property of matter more generally, 'at every level of matter as that which is potentiating or informational'. The development of a concept of affect in itself complicates Hardt and Negri's account of affective labour. For Clough, affective labour is not simply about caring for others or attempting to shape the emotional states of other human beings. Rather 'affectivity is central to the present relationship of measure and value. Whereas measure had previously provided a representation of value, now affectivity has become a means of measuring value that is itself autoaffective, producing affect in a multiplier effect across metastable levels of matter' (Clough et al. 2007: 74). Within the conceptual framework I have developed in this book, affect can be understood as the moment of encounter between the qualities generated by objects and the responses that these qualities potentialize in the bodies that experience them.

Game designers attempt to create envelopes during the experience of gameplay by manipulating the relationship between the memory, perception and anticipation of players. In addition to this, they are increasingly turning to bespoke statistical tracking systems in order to produce and manipulate affective value even when players are not directly playing games. Following a general logic of media convergence (Ip 2008, Jenkins 2008), many popular videogame franchises create applications and programs for mobile, tablet and PC devices that are designed to coexist and work alongside the games themselves. Three of these systems are the *Waypoint*, *Battlelog* and *Elite*, applications that have been released in 2011 for the *Halo*, *Battlefield* and *Call of Duty* series of FPS games respectively. In the multiplayer mode of these games, players can battle against one another individually or in teams across timed matches in varying modes (which include, for example, 'capture the flag' and 'team death match' modes). The applications provide stylized tables of graphs, maps and text that communicate various statistics about a player's performance in these games. The statistics cover players' actions in minute detail, including their accuracy for each weapon they use and the number of headshots achieved with certain weapons.

Of course, videogames have long drawn upon and utilized statistical tracking in order to operate as complex simulations. Simple games such as *Pong* displayed the player's score at the top of the screen as they played and later arcade games had a high-score table recording and denoting the best score that had been achieved on that machine (Wolf 2008, Donovan 2010). The current forms of tracking system the chapter analyses differ from older forms of statistical measure in two main ways. First, information about gameplay has become mobile. It has been exteriorized outside of the immediate game or game environment from which statistics about the game are generated and accumulated. In older arcade machines, scores were local to this or that machine, whereas systems such as *Elite* and *Battlelog* allow players to check their statistics and progression from a number of devices that are not tied to the particular console where they play the game. Second, these applications explicitly draw upon and mobilize ideas of gameplay as a form of labour. In *Battlelog* and *Elite* for example, players' overall statistics are presented as a 'Career Summary'. The notion of a gameplay career implies that the accumulation of levels, bonuses and grades in game are as, if not more, important than any enjoyment that a player derives from the gameplay itself.

Investigating these systems allows us to examine care – in particular, care for how and why one plays a game – and explore how care is shaped by a series of 'ecotechnical systems' (systems that separate out and place entities in relation to one another to shape players' habitual, perceptual and affective responses to these systems). This chapter argues that these ecotechnical systems work to shape care, as a temporal phenomenon and affective set of dispositions, by suspending and deferring negative affect. Within online videogames, an attempt to suspend and defer negative affects is related to attempts to minimize players' frustration and annoyance in order to keep the player involved and invested in the game. Designers want to discourage a player who may decide to stop playing a game, whether that be over the course of a particular gameplay session or to stop playing the game altogether. This affective suspension is important because much of the profit in online videogames is derived from selling the player expansion packs and downloadable content for the game months (and sometimes years) after a game's initial release. The games industry is economically invested in keeping players happy and engaged with the game in order to have the largest possible player base to whom it can sell the next game expansion or map pack. There is also monetary value in the statistical systems themselves. For instance, the *Waypoint* system is free to download to a mobile phone, but it requires a payment in order to unlock all the overhead maps, which are used to record the location of players' kills and deaths (I expand upon this in the third section).

Developing the concept of inorganically organized objects in relation to the statistical systems utilized in *Elite*, *Waypoint* and *Battlelog* offers a useful additive to accounts of affective labour, which posit that affective labour is simply about the production of positive states of being and emotion through the minimization of negative affects: the feeling of 'ease and well being' that Hardt and Negri describe. Rather than the production of positive feelings, affective labour under statistical tracking systems such as *Waypoint*, *Elite* and *Battlelog* can also involve the production of structures that attempt to temporally and affectively *defer* and *suspend* negative affects in ways that leave emotional situations open and unresolved. Furthermore, these systems can also be used to demonstrate how the resolution and technicity of objects that make up these tracking systems work to 'space' and 'time' entities in a way to create this affective value. That these systems explicitly position players' experience in terms of a 'career' demonstrates that the games industry does not attempt to hide or shroud this affective labour from the player through ideological techniques. Indeed, discourses of the gamer career are explicitly advertised and drawn upon as a reason in and of itself to download and use these systems.

Producing an account of the spatiality and temporality of these statistical systems also adds a layer of phenomenological experience as to how these systems are perceived through the interfaces in which they are presented. Accounts of statistical systems in new media have examined how individuals are constructed and sorted into generalized categories by technologies such as search engines (Chun 2006, Becker 2009, Amoore 2011, Cheney-Lippold 2011, Introna 2011), but have done less to examine how these processes are experienced by users themselves. The concept of ecotechnics allows us to understand how the creation of spatial, temporal and affective proximity between different entities (be they statistics in a system, or a marker on a map) can alter and shape affective dispositions to these entities.

To make these claims, the rest of the chapter forms four sections. In the first section, I develop the concept of ecotechnics from the work of Jean-Luc Nancy and theorize care from a specifically Heideggerian perspective. The next section applies these concepts through two examples that demonstrate how the *Waypoint*, *Battlelog* and *Elite* systems specifically operate to defer negative affective states and shape how one cares about the experience of playing these games. The third section contrasts these exteriorized systems with videogames such as *Demon's Souls* that track and present minimal amounts of data to the player. Rather than attempt to defer or suspend negative affect, *Demon's Souls* embraces negative experiences and, in doing so, creates immanent ecotechnical systems of care. Contrasting suspended and immanent systems of care, the final section points to the ways in which the ecotechnical systems inherent to videogames can take many forms in ways that do not have to necessarily supplement envelope power.

## The ecotechnics of care

In Nancy's work, ecotechnics refers to the way in which technology, of all kinds, operates to separate out and put beings in contact with one another. In Nancy's (2009: 89) words, 'the ecotechnical functions with technical apparatuses, to which our every part is connected... what it makes are our bodies, which it brings into the world and links to the system, thereby creating our bodies as more visible, more proliferating, more polymorphic, more compressed, more amassed and zoned than ever before'.

Technical objects work to partition bodies into their own zones and spaces, while connecting them to other bodies through these zones and spaces. For example, the telephone network allows bodies to communicate with one another but only through particular devices. Ecotechnics also 'creates space for the withdrawal of any transcendent or immanent signification' (Nancy 2009: 89) in which humans are 'exposed, body to body, edge to edge, touched and spaced, near in no longer having a common assumption, but having only the between-us of our tracings *partes extra partes*' (Nancy 2009: 91). On a concrete level, this means that technology, in all its forms, serves to partition, shape and structure our contact with other beings in the world. For example, a desk shapes how a body comports itself to write, or a mouse shapes how one accesses information on a computer. In terms of the account of objects developed in this book, ecotechnics works to transduce qualities through the way objects are arranged in order to encounter one another. Crucially, this comportment is not merely physical but also actively structures the thought processes and possibilities for thought associated with these practices.

Drawing upon Nancy's concept of ecotechnics, we can begin to refashion the term to think specifically about the forms of affective relation enabled and inhibited by the social networking and statistical systems of videogames. Ecotechnics refers to separating out entities in ways that disclose particular qualities of these entities to the senses in more or less intense or relaxed ways. Within statistical systems tracking player performance in videogames, objects are specifically ordered so as to manage players' affective experience. The processes of separation and exteriorization involved in this affective suspension are enabled through various forms of transduction. The immediacy of a particular affective encounter is spatially and temporally 'distanced' from the player through the ways in which the game allows qualities to be exteriorized, rationalized and enframed by another system or process. In doing so, qualities are transduced in players' bodies as affects. This spatialization and temporalization of affect can be understood as the ecotechnization of care.

For Heidegger, care is a basic phenomenon that structures how humans encounter the world and entities within the world. As he writes in *Being and Time*:

> Care as a primordial structural totality, lies 'before' every factical 'attitude' and 'situation' of *Dasein*, and it does so existentially *apriori*; this means that it always lies in them. So this phenomenon by no means expresses a priority of the practical attitude over the theoretical. When we ascertain something present-at-hand by merely beholding it, this activity has the character of care just as much as does a 'political action' or taking a rest and enjoying oneself.... The phenomenon of care in its totality is essentially something that cannot be torn asunder; so any attempts to trace it back to special acts or drives like willing and wishing, or urge and addiction, or to construct it out of these, will be unsuccessful. (Heidegger 1962: 238)

For Heidegger, care sets up the fundamental temporal structure of action that shapes how the world appears and what objects appear as more or less important within that world. Care encourages us to look forward and imagine possible futures and engage in activities with future outcomes in mind. These might include very basic possibilities, such as finding water to avoid death by dehydration, or more complex ones, such as completing a task at work in the hope of promotion. Care is then linked to the temporal structure of experience discussed in Chapter 4. We care about the future, which in turn sets up our activities in the present and how we reflect on past events in relation to this future and present.

Ecotechnics operate to shape the relationship between the resolution and technicity of particular objects and care. By separating out objects and distributing them in ways that are more or less accessible to humans, or more or less combinable with one another, objects are able to shape humans' ability to care about them. This influences how humans imagine and anticipate the future and shapes their present activities accordingly. For example, the relationship between a pen and a piece of paper has a particular ecotechnical relation. A person requires a piece of paper and pen if they wish to send a letter to another person. Pushing the pen against the paper with the correct pressure results in ink being released, and allows a message to be written (so long as the wielder of that pen knows how to and is able to write). The transduction of forces and material between pen, ink and paper requires that a writer places their hand at a particular position on the pen's shaft and raises their hand to a particular angle to avoid smudging the ink onto the paper. Here, the pen and paper operate as transducers in two directions. On the one hand, the ink is transduced from being a liquid state to a solid state as it adheres to the paper. On the other hand, the writer's hand is transduced by this operation: it is drawn into a particular shape, encouraged to exert a particular strength of grip and so on. The correct ecotechnical relation of proximity and distance has to be observed in order to realize the future possibility of sending a letter, which is itself based upon enabling particular transductions to occur between pen, writer and paper. Care is, then, both a temporal structure that orientates present and future activity, and a set of corporeal, affective dispositions that involve the training of particular senses, gestures and comportments to complete particular tasks.

While all objects exhibit and require ecotechnical tendencies of proximity and distance for the transduction of affects, videogame statistical tracking systems in particular work through a specific set of ecotechnical operations to separate out and quantify various affective states of care. The *Waypoint*, *Elite* and *Battlelog* systems do this by shaping how one cares about particular aspects of the game and the importance that is placed on these aspects through cutting up temporal and spatial experience into discrete markers, as well as prioritizing particular forms of sense and gesture in response to affectively stimulating experiences. This ecotechnics of care serves to translate, transport and manage the potential for affect into a feedback loop between player and game. In the case of videogames, this feedback loop is shaped by a number of different objects.

The ecotechnical systems involved in the *Waypoint*, *Elite* and *Battlelog* services involve a specific set of transducers that work to create a state of affective suspension through shaping how players care about what they are doing. This ecotechnics of care can be understood as working through a number of processes, two of which are discussed in the next section; the quantification of success and the spatialization of failure. These processes work to maintain a player's interest in a game, creating a state of affective suspension in which negative affects can be deferred or downplayed, even in the face of what can be extreme frustration, anger or despondency.

## Ecotechnics of care: Two sites of transduction

The following examples demonstrate how an ecotechnics of care operates and the ways in which care, as both a temporal phenomenon and corporeal set of affective dispositions, is shaped by the *Elite*, *Waypoint* and *Battlelog* systems.

## *Heat maps (spatialization of failure)*

The *Halo Waypoint* and *Call of Duty Elite* systems allow users to study previous matches they have played by displaying information about the match on a top-down diagram of the level. The player can cycle through various forms of information that is displayed on the map, including locations of the deaths of all the players or the individual player using the system, and a timeline displaying when these kills and deaths took place across the duration of a match. One way this information is displayed is through the use of heat markers on the map. A red sphere shows a death of the player and a green sphere shows a kill by the player. These spheres can be overlaid upon one another to form a heat map showing 'hot spots' where the majority of kills or deaths in a level take place. The denser and larger the spot, the higher the number of kills or deaths that took place in that particular area. The spatialization of the kills and deaths in a level is supplemented by a timeline and kill feed that show the time at which a particular kill or death took place and the weapon used to produce this kill or death.

Players can use this information in a variety of ways. For example, players can go back and look for areas in which they tended to be killed regularly or look to see where the majority of action tends to happen within a level. Through this process, the particular corporeal affects associated with events in a match, such as being killed a number of times, become spatialized and temporalized as a series of distinct markers on an overhead map and lines on a timeline. Playing *Halo* or *Call of Duty* online can be an incredibly intense and dizzying experience, with little time to reflect on what is happening during match play. Cutting these experiences into specific events can have the effect of allowing the player to rationalize these, often intense, emotional states into a series of codified markers that will allow them to improve their game in future. The spatial distribution of these markers on the screen is key to these systems' ecotechnical effects. Presenting kills and deaths in different colours (red and green) reorganizes and transduces the temporal and affective relationship between past events in the gameplay into tertiary retentions in these systems. The negative associations of dying are translated and embodied in the colour red and the blotches of red can form patterns on the heat maps. Displaying the kills and deaths alongside one another can produce overlaps as kills and deaths that took place at different points in the same location are displayed on top of one another. Using different colours juxtaposes these events as two different kinds of activity, even when they are displayed as inhabiting the same location. The use of colour serves to separate out and distance these activities, even when they are spatially proximate to one another. In turn, this use of colour can create different affective associations between different categories of events, which have themselves been transduced into a different visual form.

Through the ecotechnical distribution of separation and overlap, the extent to which and how players care about their performance within a match is also altered. Care can potentially take on a more abstract and detached quality. The top-down map of the level is very removed from the visceral experience of actual match play. Observing the death that occurred during a match as a red marker shapes players' capacity to care about that death, as both a previous event that really happened and as a memory of that event as experienced in the present moment. The red dots evacuate the movement that occurred between those points during match play and the player's particular attuned responses that produced these results. Furthermore, in *Elite* players cannot simply move around the map in this view. They can only use the left and right triggers on the control pad to cycle through the kills and deaths as they occurred within the match. This use of the control pad triggers is an ecotechnical partition of care in and of itself. A simple click moves time and space forwards and backwards for the player as they observe the kill and death data from the match. The distance between the points on the map and timeline is eradicated and homogenized in a single movement that betrays the sensorimotor effort required during the match.

In doing so, the map works to post-rationalize and downplay the care involved in the original gameplay. This ecotechnics of care, produced through a transduction of affect from a fraught encounter to a spatial marker, can also work to diffuse a negative affective state and encourage the player to try again, with the knowledge discerned from studying the *Waypoint* and *Elite* statistics to hand. This narrative of improvement is central to the *Waypoint*, *Elite* and *Battlelog* services. Here, improvement is not simply about an increase in the players' awareness or skill with the game, but about players' learning to manage their affective sensibilities through a suspension of negative affects associated with in game experience. Post-rationalization of the match experience using the timeline and map encourages players to focus on future improvement rather than past effort and, in doing so, to centre their current care on the immanent future in which the next match will be played.

## *Kill death ratio – (quantification of success)*

Another way in which care is partitioned is through the recording and presentation of kill death ratios. The kill death ratio is a statistic that measures the number of times the player has been killed compared to the number of enemies the player has killed. The kill death ratio is measured and presented in a number of ways within the *Waypoint*, *Elite* and *Battlelog* applications. Upon entering these applications, the player's 'career' ratio is presented. The career ratio presents the total ratio for every game the player has ever played. The player can then cycle this ratio in relation to particular game types and compare between game types. For example, a player may have a 0.96 ratio (meaning they have killed 0.96 players for every time they have been killed) on the 'team death match' game type, but a 1.45 ratio (meaning they have killed 1.45 players for every time they have been killed) on the 'free for all' game type.

In one sense, the kill death ratio does not necessarily denote the skill of a player. Success in specific game types such as capture the flag is decided by a team's total number of points in completing an objective rather than the number of times an individual player has died. However, the prominence of the kill death ratio in the statistical leader board ranking within the game and in the career statistics of the *Elite*, *Battlelog* and *Waypoint* applications can encourage players to define the quality of their play and their skill as a player in relation to this kill death statistic.

Furthermore, the player's career kill death ratio begins with the first match that the player engaged in, regardless of their skill level or experience with previous games. This can lead to the situation where players become more competent at the game over a period of time but still have a low kill death ratio because the process of learning to play the game resulted in many more deaths than kills. Having a negative kill death ratio can create a sense of frustration for the player and cause them to either quit playing, or to pursue the goal of returning from a negative to a positive kill death ratio. Although *Waypoint*, *Elite* and *Battlelog* record the match win/lose ratio, many players see this as less important than the kill death ratio. This can lead to the strange situation in which players are happy with their performance in a match even though their team lost, because they personally finished with more kills than deaths.

The kill death statistic also works to create temporal separation between players' actions and their emotional reflection and rationalization of these actions. Players may become frustrated by their play within a particular match and use the kill death ratio as a way of minimizing their frustration by relating the match to a broader narrative of trying to raise one's career kill death. Here, the statistic tracking systems work to alter the context through which human care is registered by creating temporal distance between entities and altering how the qualities of these entities are expressed to the player. Rather than focusing their care on a proximate previous match in game, the system encourages players to look forward to a more distant and abstract future goal that exists above and beyond any particular match or play session. The recording of the kill death ratio transduces the sum of particular experiences from the flux of sensorimotor action into a transportable statistic. In turn, the players' negative recollections of a particular match can appear distant to them because they are focused on a more proximate future in which they might improve and increase their kill death ratio. These examples make quite clear the ways in which the *Waypoint*, *Battlelog* and *Elite* systems work to shape players' corporeal and temporal experience of care. This ecotechnics of care works through partitioning the flow of affective responses and resonances to situations and externalizing these into the environmental metrics of the *Waypoint*, *Elite* and *Battlelog* systems.

## From suspended to immanent ecotechnical systems of care

Writing in *Gameplay Mode*, Patrick Crogan (2011: xii) suggests the simulations on which videogames are based cannot be separated from a 'military technoscientific legacy' that has had what he terms a 'profound effect on the development of computer games'. Crogan (2011: 170) suggests that all videogames operate through a shared logic or gameplay mode that revolves around the simulation of conflict, a simulation that in turn attempts to close down or foreclose contingency through a 'deterrent anticipatory mode'. In first-person shooter games, this is expressed through the ways in which 'the future is brought under control by being determined strictly in terms of the present' (Crogan 2011: 97). *Waypoint*, *Elite* and *Battelog* and the techniques these systems use to suspend negative affects could well be understood through this lens. However, not all games have to follow this military logic and other forms of ecotechnical relations of care are possible.

Games such as *Demon's Souls* point to ways in which ecotechnical systems can be used to amplify senses of positive affective relation, rather than simply suspend or defer negative affects. In *Demon's Souls*, the player controls a lone hero attempting to rid a forsaken land from demons that drive the inhabitants mad. *Demon's Souls* does not utilize an exteriorized statistic tracking application. Indeed, the whole game is designed to be incredibly opaque, providing players with little information about how they are progressing through the game and offering no formal system to reflect upon or quantify that experience. The absence of these systems creates the potential for an immanent ecotechnical system of care that is entirely different from games that utilize statistical tracking systems. For instance, if the console is connected to the Internet, the player will notice ghostly shadows moving around the environment with them. These shadows are other players, embarking on their own quests in parallel worlds on their own console. In the same way, the player will appear as a ghostly apparition to other players currently exploring the same environment. These shadows cannot directly interact or engage with the player but FromSoftware, the game's developers, has devised a number of ways in which players can indirectly contact or relate to one another.

Players can leave messages on the ground for other players to find. These messages are based upon a relatively small number of prerequisite phrases relating to the gameplay. There are three categories of messages including 'general', 'hints' and 'strategy'. A message might be 'be wary of trap ahead' or 'use divide and conquer to beat next enemies'. What differentiates this mode of relation to the types of text or voice communication in games such as *Halo* is the inherent ambiguity of a message. Players are unable to be specific when leaving a message because of the predetermined, general nature of the messages that are available to them. This can lead to surprising and novel situations. In the Boletarian Palace area, a message in front of an archway leading into a dark room may read 'beware ambush ahead'. The ecotechnical system employed in *Demon's Souls* means that there was no way of discerning any further context to this message, outside of its spatial location on the ground. Reading the message players are faced with a range of questions: Is the warning serious or a joke? What exactly will the ambush be? Is the note referring to one enemy, many enemies or possibly a trap? How much further ahead is the ambush in question? This ambiguity serves to encourage the player to reflect on the current situation, although they have no possible way of resolving the issues the note raised. If a note is useful, players can recommend the note to others. Recommending the note serves to heal the player who originally left the note in their own world. Notes with higher numbers of recommendations stay on the ground longer, providing hints for longer periods of time and presumably to larger numbers of players who travel through the space in which the note is left.

The message and recommendation system in *Demon's Souls* is an example of an ecotechnical system of care that is very different from the exteriorized statistical tracking systems utilized in *Waypoint*, *Battlelog* or *Elite*. The notes left on the ground for other players remain within the environmental context in which a player finds them. As these notes are impossible to translate outside of this context, they gain a temporal fleetingness, which negates their capacity to regulate or manage players' affective and temporal experience of care in the same way as *Waypoint*, *Elite* or *Battlelog*. The ecotechnical power of these systems is drawn from the way in which they produce fixed enduring statistics and maps that can be analysed after the event of play. This is powerful because of the ways in which these systems can orientate temporal and affective care towards the future, thus anesthetizing negative affects that may discourage players from continuing to play. Without these fixed systems, temporal and affective states of care in *Demon's Souls* are much more immanent to the gameplay experience itself. Pointing towards immanent ecotechnical systems of care highlights that ecotechnical systems in videogames are not simply about the suspension or deferral of negative affect. Immanent ecotechnical systems can structure care around the present and allow emotional states to be expressed in the present, rather than deferred or suspended indefinitely.

## The temporal deferral of negative affect

*Waypoint, Elite* and *Battlelog* are some examples of a whole range of statistical systems that are designed to suspend and defer negative affects in order to keep individuals caring about the products they buy and use. By suspending negative affects and deferring the potential for positive affects onto an open temporal horizon, these products manage affect for the explicit creation and realization of economic profit. This is quite different from how writers such as Colebrook argue affect is manipulated in videogames. As Colebrook (2014: 76) puts it:

> there is ... today, a contraction or weakening of grammars and syntaxes of cognition in the face of the instant gratification of affects. Computer games, and the cinematic and tele-visual cultural products that are inflected by game culture may have narrative and teleological components, but the dominant experience is that of intensities. A culture of shock and awe allows us to sit before a screen and enjoy the affects of horror, terror, mourning, desire, disgust, fear and excitement without sense.

Rather than providing instant gratification 'without sense', systems such as *Elite* or *Waypoint* work by attempting to defer potentially negative affects that are produced in the moment and manage affective responses to events by placing them into broader temporal and relational contexts of comparison and measure. Instead of neutralizing or shutting down negative affects, counter intuitively, these systems work by attempting to keep the potential of affect open by creating fixed and transportable statistics and diagrams of past action. Games such as *Demon's Souls*, in which care is focused on the players' present activity with no ability to go back and reflect upon previous action, can have an opposite effect in that they structure care around a present moment or encounter. The *Elite*, *Battlelog* and *Waypoint* systems are also examples of the ways in which the immaterial labour and 'playbour' (Kücklich 2005) of videogames is being extended outside of the actual event of gameplay. The iPhone and Android applications for these services allow mobile players to monitor their progress at any time. In doing so, affective suspension and deferral is made possible away from the interface environment or envelope that emerges during gameplay.

The ecotechnical systems at work in videogames are also expanding to other products and services. For example, Audible, an online audio book company owned by Amazon has an application with a statistics system embedded in it. The app allows listeners to track how much time they have spent listening and users can also unlock specific badges depending on how, where, when and how long they listen. As such, the ecotechnics of care evidenced in videogames is a small example of a much larger process in which environments are being designed in order to alter the relationship between consumers and the services and products they buy and use. While the Audible application might be designed to create different forms of affective relation from the systems associated with videogames, the forms of statistics tracking systems employed are similar in the sense that they are actively attempting to shape the temporal and affective structure of care. Pointing to how various deferred and immanent ecotechnical systems operate provides a vocabulary to think about the implications of these systems and their potential impacts on how people's future activities are organized and shaped by the companies that create these systems. The next chapter investigates this issue by focusing on how the logics and techniques of envelope power utilized in videogames are spreading to other forms of non-gaming interface. In doing so, the chapter offers suggestions as to how to interrogate the logics of the interface envelope and the potential problems that envelope power creates for perception, critical thought and decision-making.

# 7
# Envelope Life

> [W]e are enveloped, entangled, surrounded; we are never outside without having recreated another more artificial, more fragile, more engineered envelope. We move from envelopes to envelopes, from folds to folds, never from one private sphere to the Great Outside.
>
> LATOUR (2011: 158)

As Latour suggests, most populated environments on earth are loaded with technical objects to the point of ubiquity. One cannot simply step outside of, escape or exit technical envelopes of light, heat, shelter, medicine and so on, because these envelopes enable human life to continue. In Latour's (2011: 158) words, 'Naked humans are as rare as naked cosmonauts. To define humans is to define ... the life support systems ... that make it possible for them to breathe'. While Latour's definition of an envelope differs from my own, the same logic can be used to speculate about how the envelopes generated by modern videogames are spreading to other technologies, interfaces and types of activity. In the supposedly 'gamified' contemporary world (Frith 2013, Zichermann and Linder 2013, Raessens 2014), putting the videogame control pad or stick down does not mean simply leaving an envelope, but potentially moving from the site of one possible envelope to another.

To investigate these claims, the chapter moves through four sections. The first section discusses the rise of discourses of gamification and how processes of gamification attempt to modify non-game environments to potentially produce interface envelopes. The next section turns to three specific examples of gamification in action and discusses the extent to which they generate envelopes and envelope power. The third section reflects on the potential of interface envelopes for society and points to how the logic of the interface envelope (which acts to shape the relationship between anticipation and memory) might be critiqued, called into question or rethought to create alternative modes of being in the present. Finally, the chapter cautions against a wholesale rejection of the logics of the interface envelope and points to its positive pharmacological potential.

## Gamification

The suggestion that the logics implicit in videogame design are infecting other arenas of life under late Western capitalism is not new. Mackenzie Wark (2009: 6) suggests that 'the real world appears as a video arcadia divided into many and varied games. Work is a rat race. Politics is a horse race. The economy is a casino'. Wark sees the gamification of life as predicated on a logic of targeting and based around a binary of win-lose situations. In his words, 'everything has value only when ranked against something else; everyone has value only when ranked against someone else. Every situation is win-lose, unless it is win-win – a situation where players are free to collaborate only because they seek prizes in different games' (Wark 2009: 6). Here, 'play is no longer a counter to work. Play becomes work; work becomes play' (Wark 2009: 16). Like Crogan (2011), Wark is keen to link the logics of what he calls game-space to previous developments in computing technology that emerged from the military industrial complex. In both Wark and Crogan's account, the game worker/player/subject's capacity to affect is based around and predicated on the notion of the target: 'what kind of being is a gamer? One who comes into existence through the act of targeting. To target is to isolate something against the dense, tense fibres of the network, maybe to destroy it, but always to assign it a unique value' (Wark 2009: 149).

Videogames can now also be seen as spheres of work in their own right (Terranova 2000, Malaby 2006, Millington 2009, Tobias 2010, Munster 2011, Karatzogianni and Junstman 2012, Valiaho 2012). On a governmental level, Kucklich (2009: 344) argues that videogames are social factories that 'appear...to fulfill primarily an economic function but which also create...social and cultural capital, as well as forms of political organization, which in turn feed back into the business models of the providers of virtual worlds'. In a similar vein, Yee (2009: 70) argues that 'video games train us to become more industrious gameworkers'. Yee suggests that videogames often involve repetitive tasks that mimic work practices under late capitalism and that the line between play and work has become increasingly blurred. Kucklich (2005) argues that this form of control does not simply occur within the games themselves, but also through parasocial practices such as game modification and forum use (also see Postigo 2003).

While Wark (2009), Crogan (2011) and other writers such as von Hilgers (2012) argue that the logics of videogames are implicitly carrying over to everyday life, product and interface designers are now explicitly adding videogame-like mechanics to a range of non-videogame activities. Zichermann and Cunningham (2011) term this process 'gamification', which Raessens (2014) understands as the ludification of culture. Zichermann and Cunningham (2011: xiv) define gamification as using 'the process of game-thinking and game mechanics to engage users and solve problems'. In practice, applying gamification techniques to non-game software and products involves incorporating a number of game mechanics into the interfaces of these objects (such as in the Audible app discussed at the end of the previous chapter). Deterding et al. (2011: 4) outline five of these possible mechanics:

1. Interface design patterns such as badges, levels, or leaderboards.... 2. Game design patterns...or game mechanics.... 3. Design principles or heuristics: guidelines for approaching a design problem or evaluating a design solution. 4. Conceptual models of game design units, such as the MDA [Mechanics, Dynamics and Aesthetics]...framework.... 5. Game design methods, including game design specific practices such as playtesting and design processes like playcentric design or value conscious game design.

The logic behind the rise of gamification, according to Deterding et al. (2011: 1), is that 'video games are explicitly designed for entertainment rather than utility – they can demonstrably produce states of desirable experience, and motivate users to remain engaged in an activity with unparalleled intensity and duration. Thus, game design is a valuable approach for making non-game products, services, or applications, more enjoyable, motivating, and/or engaging to use'. Examples of gamification include locative applications, such as FourSquare, which encourage the user to check in at different locations and award badges for doing so (Wilson 2012). These badges include 'Local' (awarded when a user checks into the same place three times in one week) and 'School Night' (awarded when the user checks into a place after 3 a.m. on a weekday). Users can also compete with others to become 'mayor' of a particular place by checking in more than anyone else over the period of sixty days.

Other examples of gamification include Grush, a portmanteau of gaming toothbrush. This is a smart toothbrush that syncs to smartphones in order to encourage children to brush their teeth thoroughly (LeJacq 2014). A mount connects the smartphone to a mirror and children can play different games using the brush as a controller that sends motion data to the phone. In the game 'Monster Chase', for example, the child's mouth is filled with monsters that can only be defeated by brushing over them. Furthermore, parents can monitor children's brushing using a dashboard application and even send this data to their dentist for further review. In short, gamification is both a process of applying game-like mechanics to non-game activities, as well as utilizing game-like aesthetics in non-game products.

While there are examples of civic gamification (Feffer 2011), the enthusiasm for gamification has largely emerged from the world of business, where the explicit purpose is to use game techniques to produce new products, increase user participation, create stronger loyalty between customers and companies and ultimately increase the profits of companies that employ and incorporate these techniques (Zichermann and Linder 2013). Alongside this, critiques of gamification are emerging (Bogost 2011) that seek to go beyond instrumental issues of how to better imbue game mechanics into non-game systems and focus on the wider implications of these processes in generating particular structures of perception or subjectivity (Fuchs et al. 2014). Indeed, following the arguments developed across the previous chapters, we could see gamification as one way that the logics of videogame interface envelopes and envelope power are extending into more and more aspects of everyday life.

## Non-gaming interface envelopes

The rhetoric surrounding gamification clearly shows how companies are attempting to spread the logics of the interface envelope and in turn envelope power to other forms of digital interfaces outside of gaming. Of course, the implications of these envelopes are not uniform. Nonetheless, the examples I develop in this section (of augmented reality mapping applications, the Nike Fuelband, and smart cars and meters) show some commonalities in the potential impacts these envelopes are having on the organization and emergence of perception and subjectivity within a world that is increasingly accessed by digital interfaces.

### *Augmented reality mapping*

The recent development of augmented reality applications for smartphones shares a number of key factors with interface design in videogames (Lammes 2008, Chesher 2012, Verhoeff 2012, Katz 2013). On the most simple level, there is a key overlap in the way that videogames and augmented reality present environmental data to the user. Information is locative and placed as a translucent overlay in front of the object that the information describes or refers to. As Jørgensen (2013: 162) puts it, 'augmented reality technology resembles the most common interface design trend in digital games: using overlays and other kinds of emphasis that call attention to specific information in the... environment'.

Understood as a further development of previous GPS and locative mapping applications, augmented reality map applications (such as *Yelp Monocle*, *Theodolite* or *lookator*) work as forms of overlay to 'bind... components occurring simultaneously together on flat screens as diagrammatic networks' (Neidich 2014: 266), which 'increase mental efficiency' because they 'can off load operations that would normally go on inside our heads upon external props, thus decreasing the energy requirements of the brain as well as freeing up neural processes to do other things' (Neidich 2014: 271). *Yelp Monocle* places names and reviews of businesses on top of an image of a space when it is viewed through a smartphone camera. *Lookator* works in a similar way to show the location of Wi-Fi hotspots. *Theodolite* is a more sophisticated viewfinder for displaying topographical data. When looking through a device's camera, *Theodolite* overlays information such as bearing, heading, altitude, and elevation as well as allowing users to set custom map points that continually update heading and distance relative to their current location.

Using apps such as *Lookator*, tertiary retentions are exteriorized and spatialized in a dynamic way onto the environment. Rather than a fixed point of a map that provides a static overview of a space, markers that denote the location of Wi-Fi hotspots become seemingly fixed to the objects in the world themselves. By fixing information onto objects, these interfaces attempt to narrow the gap between representation and object but, in doing so, also push users' perception into a narrowly defined present tense. Using apps such as *Theodolite* or *Lookator* can focus attention on the now and the perceptual present of exploration as the interface dynamically responds to the device's spatial positioning and coordinates. These apps potentially short cut the relationship between tertiary and secondary retention, by creating and presenting tertiary retentions to the user that stand for or replace the need for secondary retention. Having markers placed in the field of view means that users have less requirement to rely on secondary retentions. At the same time, the heads-up display interface minimizes the requirement for users to anticipate or project their thought too far into the future towards the next potential decision point because the interface constantly adjusts to the users' position in space.

Writers such as Stiegler and Neidich decry the rise of GPS and mobile mapping applications because they argue these technologies ultimately lead to the creation of incapacitated bodies. In Stiegler's (2007: n.p.) account, the psychopower of GPS ultimately leads to a situation in which:

> we lose our sensori-motor schema formalized by the system as it becomes automatic. The more we delegate the execution of series of small tasks that make up the warp and woof of our lives to the apparatuses and services of modern industry, the more vain we become: the more we lose not only our know-how but our know-how-to-live-well: the only thing left for us is to consume blindly.

While Neidich (2014: 283) recognizes that the 'politics of cognitive capitalism...produce neuromodulatory consequences, which can expand or contract the power of the minds eye', in the case of GPS, he focuses on the ways that it contracts the capacity of the brain, leading, as I discussed in Chapter 5, to a 'capturing and externalization of thought at the expense of deep internalized contemplation' (Neidich 2014: 281).

Understanding augmented reality apps as an expression of envelope power, rather than neuropower or psychopower, leads to a different emphasis from either Neidich or Stiegler's account of GPS. Considered as an expression of envelope power, an app such as *Theodolite* does not simply work to reduce capacities for thought, but actively produces new habitual and synaptic relations between body and brain. The resolution and technicity of the objects that make up the map interface cultivate new modes of perception and use, while training the user to become reliant on the interface. For example, using *Theodolite* provides information to the user that they would usually be unaware of, such as their altitude or elevation. In this case, rather than being 'incarcerated' by the mapping application, the user becomes more actively sensitised to their own spatial location and information associated with this location. However, instead of empowering the individual, this increased sensitisation may lead to the user feeling as if they have to rely on *Theodolite* more deeply *because* they are sensitised to minute temporal and spatial distinctions that the app cultivates.

While in its infancy, the rise of wearable technology such as Google Glass points towards a possible future in which augmented reality becomes an ordinary part of human vision, rather than an interface that users can only access by holding a smartphone or tablet in front of them. In this case, Google Glass in concert with augmented reality mapping may successfully create envelope power through the way the device and apps associated with the device both trains users to develop new sensitivities and capacities to experience the environments they move through, while also encouraging dependence on these devices because they cater to such sensitivities and capacities.

## *Nike FuelBand*

Quantified self devices, such as the Nike FuelBand, are another example of wearable technologies that are designed to be kept on the body at all times and which also have the potential to generate an interface envelope. The FuelBand is a bracelet worn around the wrist with an embedded accelerometer and wireless synchronization. The accelerometer records movement and can communicate with a smartphone. The FuelBand converts movement into a universal unit of measurement Nike calls NikeFuel. Nike does not disclose what the unit of measurement 'NikeFuel' actually is (although the band also records and displays calories burnt, number of steps taken and distance travelled as well). Using the smartphone application, users can set a daily goal for how active they want to be and as they move 'fuel' is added to the total, which is represented both numerically and through a row of light dots on the device that move from right to left and from red to green. Users can view their progress by checking their smartphone, or by pressing a button on the device itself. These two ways of interacting with the FuelBand shape the limits around which protention and retention potentially become organized. For example, users can activate a 'win the hour' feature. This feature asks the user to 'get moving for 5 mins in a row each hour'. There are also daily and weekly goals as well as streaks, which the user activates if they manage to break the record for each day's Fuel count for more than three days in a row.

Using the Nike FuelBand, an envelope potentially emerges through the relationship between intensive bodily movement, extensive movement across or through space and the FuelBand itself. The device potentially creates an envelope by transducing encounters between a body and the Nike FuelBand to create a new object: NikeFuel, which has its own technicity and resolution when it is displayed on either the Nike FuelBand or on the smartphone app that is used in conjunction with the band. When wearing a Nike FuelBand, bodily movement forms the basis around which 'the now' comes to be structured as walking, stretching or jumping is converted by the device into the quantified unit of NikeFuel. In doing so, the Nike FuelBand transduces the relations between body and device to create objects whose qualities appear to create positive affects for the person using the device.

These objects include the row of dots that fill as the user moves towards their goal. The dots on Nike FuelBand are designed to appear in high resolution in order to create affects of progression by generating synaesthetic links between body, brain and device. The dots appear through the translucent surface of the band and as the user earns more fuel, the dots change colour from red to yellow to green. Colours have a variety of affective associations that cross between cultures (Adams and Osgood 1973). Within the United Kingdom, for example, red, amber (a kind of orange-yellow) and green have particular associations with traffic light systems. Red is usually used to denote a warning to stop, amber to 'get ready' (to either stop or go) and green to go. More generally, red is also associated with danger or considered a broadly active colour, while green is associated with positive affects, such as health and vitality (Grossman and Wisenblit 1999). In drawing upon these kinds of cultural association, the colours on the band attempt to create a relation between bodily movement and previous habitually inscribed and implicit understandings about safety and danger. In other words, the FuelBand displays these colours to encourage the user to continue to move and use the device, even if they don't explicitly or consciously link the colours on the band to the cultural affects that prompt this movement and use.

The lights on the FuelBand also generate a technicity by conforming to the logic of temporal homogenization discussed in Chapter 4. By splitting intensive movement into a series of quantitative, spatially equalized blocks on the band itself, bodily movement becomes present as a series of units to be filled. In checking the FuelBand and anticipating when the next dot (or whole row of dots) might be filled, the user and FuelBand potentially form an interface envelope in which the now appears through a relationship between a narrowly spaced set of tertiary retentions (the current amount of NikeFuel gained) and a protention that anticipates increasing this amount until a goal is achieved.

Whereas maps on augmented reality apps potentially produce an envelope through the way they offer information in relation to a user's field of vision, Nike FuelBand creates a broader envelope organized around larger units of time, such as hours, days and weeks. Nonetheless, the logics that shape the design of both technologies are similar. Both create tertiary retentions that are designed to shape the emergence of the now by altering the users' protentions (their anticipation) around units that encourage the continued use and engagement with these products.

## *Electric cars/smart meters*

The logics of the interface envelope are also apparent in the development of ECO indicators in electric cars such as the Nissan Leaf or Toyota Prius.

In the case of the Nissan Leaf, software constantly records a number of variables including accelerator pedal operation, brake pedal operation, driving conditions, traffic conditions, heater and air conditioner usage and the amount of time the vehicle is not moving while the vehicle is in its ready-to-drive mode. This information is then communicated to the driver on a small eyebrow display. The more efficiently the user drives, perhaps by applying the brake pedal lightly, the more a semicircular gauge on the eyebrow display fills. Once full, a tree appears next to the gauge. The more efficiently the user drives, the more trees are added to the display. In this way, the interface is explicitly designed to draw the driver's attention around the now, broken down into a series of micro-movements and gestures as the driver attempts to transduce bodily movement into new objects or tertiary retentions (the additional trees displayed on the eyebrow display).

By encouraging the driver to create new trees on the eyebrow display, the resolution and technicity of the ECO system in the Leaf short cuts the relationship between protection and secondary retention and in turn potentially creates an envelope. Rather than observing the environment the driver is passing through, the driver may be concentrating on exactly how hard they are pushing the accelerator, or trying to get a feel for the gradient of the road to see if they can coast, save energy and thus fill the dashboard bar (on the car as an extension of the body, see Katz 1999). In doing so, the now is organized around a constant anticipation enabled by a tertiary retention (seeing the bars on the dashboard fill) in an attempt to create some affective sense of satisfaction that 'energy efficiency' has been achieved.

ECO meters explicitly work to increase their users' capacity to sense difference, by quantifying and breaking down usually unknown or ready-to-hand variables that contribute to the running of particular technologies. While these technologies are often praised for encouraging forms of environmental awareness and energy efficiency, the account of the interface envelope I have been developing would suggest an alternative narrative. Although it may be true that these forms of envelope produce energy efficiency, they also potentially bring different forms of subjectivity into being. These forms of subjectivity may be categorized as a kind of mild, albeit constant, anxiety, in which an individual's protentions are organized around a constant desire to check, compare and compete with others using the same systems. Furthermore, if tertiary retentions are used to short cut the relationship between protentions and secondary retentions, this can create a situation in which it is difficult to think creatively about a situation or deviate from the possible future that a technology is trying to bring into being, precisely because the user's protention is organized around a now that is always on the cusp of emerging and that always requires an immediate response from the user.

One might go as far as saying that these envelopes actually damage and curtail the possibility of dealing with the problems these technologies are supposed to address. For example, ECO meters work to individualize problems such as energy efficiency and encourage a kind of short-termist, reactionary form of thinking. Rather than reflecting on the state of the energy industry and its over reliance on fossil fuels, users are perhaps trained, through the emergence of these envelopes, to treat energy efficiency as their own personal problem, for which careful monitoring is the only solution. Furthermore, organizing perception around a now that is subject to micro-differentiation makes it very difficult to organize or generate a shared consciousness that would allow people to work together to actually deal with the underlying causes of these problems. The Nissan Leaf's ECO mode is designed to create efficiency through direct competition. When connected to the Internet, the car can rank the driver's efficiency compared to other Nissan Leaf owners from around the world. Here, the goal of energy saving is channelled via a logic of opposition (no matter how friendly) in which multiple bodies interact at a distance, but only via the envelopes that make this connection possible. Very similar to multiplayer games such as *Call of Duty* discussed in previous chapters, the ecotechnics of these relations does not cultivate or encourage engagement beyond the fleeting present and, thus, might minimize the potential for new forms of collective politics to emerge.

Furthermore, the logics inherent to these envelopes are not limited to electric cars. Digital smart meters are now being introduced by energy companies to encourage people to track and reduce their home energy expenditure. These meters offer real-time data collection of energy use, which allows companies to bill users each month for the exact amount of energy they use. Like the Nissan Leaf, these forms of data collection have also created the opportunity for mobile and web-based apps to be developed that allow users to compare and compete with the energy use of their neighbours. While these systems are still in their infancy and are based upon comparisons between monthly slices of data, rather than real-time updates, apps such as Opower's 'Social Energy' for Facebook are designed to encourage users to compete with others in their area to see who can use the least energy. More problematically than the ECO meters in electric cars, attempting to be as efficient as possible with smart meters can encourage behaviour that has negative bodily effects on those subject to smart meters. As Pegoraro (2012) reports, even without real-time monitoring and comparison, a smart energy meter user admits: 'Last winter my bills were around $70', he said. 'This winter, I think I haven't passed $30 because I'm getting into this strange competitive mode and I'm willing to live in the cold'. The envelopes potentially generated by these kinds of interface can have real corporeal effects and so should be taken seriously as a site at which new forms of subjectivity and understandings of the world can be brought into being.

## Questioning envelope life

Looking outside of games demonstrates that other forms of interface are now being designed that attempt to generate envelopes and create envelope power by cutting the now into smaller and smaller segments for the specific purpose of creating more profit for those who own and distribute these interfaces. In relation to electric cars and smart meters, these logics could be linked to emerging forms of neoliberal subjectivity (Harvey 2005, Gill 2008, Read 2009), where profit is generated through encouraging lack, austerity and thriftiness on the part of the user of that interface. Understanding these forms of interface as part of what could be termed an 'envelope economy', rather than an attention or retentional economy, provides an alternative diagnosis about what these systems potentially do as well as where any critiques can be directed.

Opponents of cognitive capitalism and the attention economy, such as Rushkoff (2013), Stiegler (2007, 2009b) and Hayles (2007, 2012), might argue that the games and services I have been discussing encourage present shock or short attention span. But, they also propose a number of solutions to the problems that an attention economy brings. For Rushkoff (2013), present shock can be avoided by simply slowing down and disengaging from forms of digital technology that constantly emphasize the importance of the present moment above and beyond everything else. In his words, 'when things begin accelerating wildly out of control, sometimes patience is the only answer. Press pause' (Rushkoff 2013: 8). Stiegler (2013b) presents a more technical solution, albeit one based on traditional values of education and improved intergenerational communication. As he argues, going against the psychopower of attention capture and the malaise this generates requires a 'common mnemotechnical culture' based upon 'writing as the literal/ literary synthesis of the flux of (past) consciousness' in order to create a generation of individuals who can 'revisit the totality of rational knowledge through the time-contraction device of the We (the condition of adoption) forged in and by this writing' (Stiegler 2011b: 150). Hayles suggests that the 'new media' which encourage hyper attention can also be used to inventively foster modes of deep attention. Illustrating this point, Hayles (2007: 197) gives the example of interactive fiction such as *Galatea*, which requires players to interact with AI in sufficient depth in order for the story to continue to unfold.

Different from either Stiegler, Rushkoff or Hayles, developing an account of the interface envelope and envelope power raises both an altered set of critiques and an alternate form of proposed interventions into the problems and promises of an attention economy. The games and services discussed in this book are not concerned with producing an incapacity to pay attention for long periods of time, but are precisely about an amplification and opening of the now as a particular mode of temporal and spatial differentiation. Envelope power is designed to increase individuals' sensory sensitivity, embodied as a fine-tuned capacity to sense difference and discriminate between ever more narrowly spaced and timed entities. If this is the case, then it is vital that we consider how we respond to an envelope power that seeks to organize spatial and temporal perception around a continuously modulating perceptual present. In the next section, I present three possible responses. The first suggests a need to retain a sense of the future as unknown and outside of logics of anticipation, the second argues for a rejection of the instrumental logic of the interface envelope, and the third points to the potential creation of new horizons of secondary collective protentions, both within and outside of videogames.

## *Retain a sense of the future as unknown*

First, one can emphasize the need to retain a sense of the future as unknown and outside the modulatory capacities of interface envelopes. This is an important point, because as Stiegler (2011a: 45) argues:

> the reduction of trust (and of time, that is, of belief in a future) to pure calculation, which would be capable therefore of eliminating everything incalculable, is what radically destroys all trust, because it destroys all possibility of believing: all possibility of believing in the indetermination of the future, in the future as indeterminate and in this indetermination as a chance, an opening to the future as to its improbability.

For Stiegler, the forestalling of the future via technology is central to controlling people who use that technology because it diminishes their ability to think about the world as being anything other than what the 'culture industries' that create the technologies of the interface envelope suggest it should be. However, as I have argued across chapters three, four and five, interface envelopes work via a logic that does not simply try to forestall possible futures from coming into existence via processes of calculation, but instead actively works to 'incorporate the very unknowability and profound uncertainty of the future into immanent decision' (Amoore 2013: 9) by encouraging and cultivating embodied skill in players. In this case, a response to envelope power is not a matter of simply trying to avoid the forestalling of the future via the options that digital interfaces offer. The resolution and technicity of objects in an interface environment are modulated precisely to incorporate the contingency of players' future action into their options, which in turn gives the player a powerful sense of control and agency, while drawing them ever closer to the narrowly spaced retentions and protentions of the envelope.

One solution to the problem of envelope powers capacity to anticipate futurity might be to make envelopes and the modulations that constitute them more plastic, malleable or bi-directional. For the most part, in the games and systems we have examined, modulation is largely a one-way process. While the resolution and technicity of objects may modulate dynamically according to sets of objects such as algorithms and rules running in the background, their capacity for modulation is normally predetermined by the designers of those objects. If users could modulate the resolution and technicity of objects in an environment according to their own goals and intentions, they could create broader envelopes in which retentions and protentions are less narrowly differentiated from one another. This could open spaces in which potential futures were not anticipated from within the logics of interface design and publishing.

However, examining systems in which such modification is possible do not offer encouraging signs. For example, the massively multiplayer online role-playing game *World of Warcraft* allows players to heavily modify the user interface and add player-developed tools to aid all manner of tasks in game, such as a threat meter to anticipate when an enemy will perform a powerful attack (Kow and Nardi 2010). As Cockshut (2012) argues, modifications can allow users to play more effectively and give them a greater sense of ownership over the play experience. But, ironically, this user modification also serves to pull players into a more narrowed envelope of experience. Drawing upon an ethnography of *World of Warcraft* players, Cockshut (2012: 222) suggests the use of interface modifications is about attempting to fulfil the players' desire 'to not...miss any key information' even if this means the screen becomes so filled with icons the player can see little of the surrounding environment. In *World of Warcraft*, adding more information on-screen serves to increase the resolution of objects in the environment by disclosing more of their qualities in the form of numerical statistics and timers. At the same time, these modifications also exteriorize players' secondary retentions (subjective memory) into tertiary retentions. This creates a situation in which players rely upon and practise their secondary retentions less and less. In doing so, the technicity of these modifications serves to organize players' perception around a very narrowly spaced set of retentions (the icons on-screen) and protentions (the possibilities those icons enable). While enabling greater efficiency for the player embarking on group raids or battles, in the case of *World of Warcraft*, the ability to create new objects to engage with the interface environment only seems to amplify the interface envelope the developers wanted to generate.

A more robust way to maintain a sense of the openness of the future is to try to avoid exteriorizing secondary retentions into tertiary retentions, which are key to the emergence of interface envelopes at all. Central to a number of the games and systems I have analysed, in particular *Call of Duty* and *Final Fantasy XIII*, is the quantification of the play experience and clear visual representations of these forms of quantification displayed prominently within the interface. This information is crucial to the creation of interface envelopes because it minimizes the need for the player to think or reflect too closely on immediate past experience within the game. In many FPS games, rather than remembering how many shots they have fired and if they need to reload, the player can simply look at the ammunition counter, which is prominently displayed in the corner of the screen. This means players can focus on the immediacy of action and anticipate what is to come, rather than dwell on a past that may have expired minutes ago.

Counterintuitively, it is the space of subjective memory (secondary retention) rather than anticipation (protention) itself that is key to undoing the logics of anticipation that govern envelopes. Rather than using tertiary memory as a crutch, users could try to learn to retain information as secondary retentions. Within videogames, the advantage of such a tactic is clear. By memorizing routes of enemies, maps and other patterns, players can subvert the intentions of the designers. This is most evident in what is known within gaming communities as 'speed runs', where players who meticulously study the minutiae of game environments and mechanics expose bugs and flaws that allow them to complete these games in very short time frames. These speed runs are only possible by not remaining in the interface envelope and actively attempting to step outside and broaden the boundaries that constitute its limit.

Outside of gaming, one can think of a similar response to technologies such as smart metering. While it may seem helpful for the user to continually rely upon the tertiary retentions transmitted to the smart meter as a way of tracking energy usage, as I discussed in Chapter 4, this is actually a form of hypomnesis (Stiegler 2007). To break a cycle of control based on hypomnesis, users of technologies such as smart metering or ECO meters could create broadened interface envelopes by altering the temporal unit of comparison they use to make sense of their energy consumption.

Rather than comparing differences in energy usage over minutes or hours, they could instead compare energy usage over days, weeks or months. This very simple technique creates an altered ecotechnical relation with the objects that make up the smart-meter technology and alters the relationship between anticipation, memory and perception, which in turn disrupts the interface envelope the designers of the smart meter might be attempting to generate. Rather than appearing as a micro-differentiated point that needs to be constantly anticipated and attended to, the now of energy usage becomes situated within a broader temporal horizon, which enables the user to think more critically about their consumption.

## *Reject the instrumental logic of the interface envelope*

Second, responding to enveloped life requires a rejection of the instrumental and individualizing logic of the interface envelope. As emphasized in Chapter 5, envelope power is not about *eliminating* differences between individuals' primary and secondary retentions. While envelope power does attempt to cultivate different retentions and capacities, it still seeks to do so for instrumental means. For the videogames industry, envelope power is an extremely effective means of maximizing the profit derived from particular products.

However, users can recalibrate the technicity and resolution of objects in ways that suit their own purposes, rather than those of the games industry. This potentially opens up new questions about what is valuable in videogames and allows users to play a greater role in defining this. This is particularly the case in sandbox games such as *Garry's mod*. *Garry's mod* is a popular game based on Valve's Source game engine. *Garry's mod* gives players a tool kit with which to experiment as they wish and even create their own games. This has led to a variety of popular player-created games, such as *Trouble in Terrorist Town* and *Prop Hunt 2*. Rather than entering into the more or less linear experiences and storylines of single-player games such as *Tomb Raider* or even open world games such as *Red Dead Redemption* or *Grand Theft Auto V*, here players can modulate the resolution and technicity of objects in the environment for their own purposes. The nature of the modification scene means that games are constantly being created for *Garry's mod* and the tools used to create these games are simple and easy for non-programmers to use.

Games such as *Trouble in Terrorist Town* and *Prop Hunt* are multiplayer, but they also require and encourage teamwork and communication. Indeed, key elements of both of these games revolve around the extra-diagetic discussion over voice or text communication, rather than tertiary retentions that are embedded in the game environment. In *Trouble in Terrorist Town*, players try to convince others that they are not the traitor, while in *Prop Hunt*, the hidden player, with little else to do other than hide, can mock or call out other players as they try to find them. What is interesting about *Trouble in Terrorist Town* or *Prop Hunt* is that the time limit for each match is often long and playing these games can involve spells of being inactive or wandering more or less aimlessly through the environment. The interface environments of these games are then clearly designed to facilitate or enable forms of sociality or being together that are generally discouraged by the envelopes in casual online play in multiplayer games like *Call of Duty*. Crucially, it is the creator-users that define and redefine these forms of sociality and the purposes they serve within the game.

Although videogames are forms of mass produced, identical, tertiary retentions, modifying and creating games points to the potential for actively encouraging alternative forms of sociality which can serve a wide range of purposes beyond the economic motives of the games industry. Modifying games to encourage forms of play based on sociality and communication in the first instance is, then, one way to question or respond to the logics of the interface envelope encouraged by AAA videogames.

## *Create new horizons of secondary collective protentions*

Thirdly and finally, a response to envelope life might involve the creation of new forms of secondary collective protentions that work to limit the efficiency of envelopes that attempt to individualize users around a continuously modulating perpetual present tense. Stiegler (2011a: 113) defines collective secondary protections as 'horizons of expectation common to a group'. These expectations include shared forms of somatic behaviour among a particular culture, such as ways of walking or standing, as well as appropriate forms of social conduct and language use. Videogames do create shared forms of somatic collective secondary protection, such as the habitual skills developed in *Call of Duty* or *Street Fighter*. More often than not, these skills are used to gain personal victory, even in group games and situations. Even in multiplayer games that supposedly require teamwork, such as *Call of Duty*, *Halo* or *Battlefield*, interface envelopes and the associated apps that support them often work to individualize players' actions. As Chesney et al. (2014) argue, using a very different vocabulary, the more somatic secondary retentions FPS players develop, the less likely they are to cooperate with other players on the same team. In other words, the more skillful they are, the less likely they are to help other players in pursuit of their own individual goals. Developing shared kinds of collective protention is important because protention and retention are intimately linked. By creating new forms of collective anticipation and expectation that are not shaped via the individualizing logics of the interface envelope, users can develop a more critical capacity to selectively engage in ways that are rewarding for them rather than the designer or company that created the interface environment. In doing so, players can make space for more diverse forms of being and thinking.

Using interfaces to create collective secondary protentions might involve the building of new communal worlds in which envelopes are not as individualized as they are in many big-budget videogames. This may sound idealistic or utopian, but these kinds of world are already coming into existence. Many of these worlds are being built using open source software and graphics engines to create alternate kinds of videogames and gameplay experiences. For example, the massively popular *Minecraft* series provides a system in which players can explore and create their own worlds constructed out of randomly generated simple-textured cubes. The simplicity of the cube environments of *Minecraft* means that the computer-generated maps are near infinite in size.[1] Players can explore these environments in three modes. The first is a survival, or story mode, where the player is cast into a randomly generated environment and has to mine resources to survive and ultimately access a region called 'The End' where they must destroy an Ender Dragon to complete the game. The second is a creative mode, which removes several restrictions around damage and allows the avatar to fly. In this mode, players can simply build structures and objects that they desire. A multiplayer mode allows players to create their own servers and invite other people to play in a shared world where everyone involved can create and build structures. The graphically simplistic cubes the player encounters form the basic ontological unit of *Minecraft*'s world. While different cubes have different capacities, they can all be potentially

---

[1] In a note on his blog, the creator of *Minecraft*, Markus Persson (2011 n.p.), explains that the maps are not truly infinite, because the software that generates the maps is limited by the hardware of the computer. As he explains:

> First of all, let me clarify some things about the "infinite" maps: They're not infinite, but there's no hard limit either. It'll just get buggier and buggier the further out you are. Terrain is generated, saved and loaded, and (kind of) rendered in chunks of 16*16*128 blocks. These chunks have an offset value that is a 32 bit integer roughly in the range negative two billion to positive two billion. If you go outside that range (about 25 % of the distance from where you are now to the sun), loading and saving chunks will start overwriting old chunks. At a 16/th of that distance, things that use integers for block positions, such as using items and pathfinding, will start overflowing and acting weird. Those are the two "hard" limits.

broken apart and connected to one another by players. Unlike the glossy environments of AAA games that carefully try to hide their artifice from the player, *Minecraft* revels in its artificiality. Indeed, the clear exposure of the game's artificial, constructed nature is key to its capacity to generate new horizons of protention.

As *Minecraft* has become more popular, groups or teams of players have begun to work together to build their own structures and worlds. One example would be an ongoing project to build the fictional world of George RR Martin's 'A Song of Fire and Ice' out of *Minecraft* cubes, called WesterosCraft. At the time of writing, this world is about 800 square kilometres, equivalent to the size of Los Angeles. In worlds such as these, there is no clear goal and no clear mechanics around which player action is organized. Rather, the simplicity of the game environment invites and encourages forms of building that require active and broad ranges of protention, which have the potential to become collectively organized. In WesterosCraft, this includes the self-organization by volunteer builders, coders and moderators, all of whom have worked together to create the WesterosCraft map. To achieve this, they draw upon multiple forms of skill and knowledge, many of which lie outside of the necessary skills to simply manipulate the *Minecraft* interface.

Furthermore, the servers on which these maps are built are not run or owned by the studio that created *Minecraft*. This means that they remain outside of their direct control, unlike the peer-to-peer connections on which many console multiplayer games such as *Call of Duty* operate. The key difference between games such as *Minecraft* and *Call of Duty* is not that creating alternative horizons of protention in *Call of Duty* is impossible. People can, and do, create machinima, or hack the levels of the game (Lowood 2006, 2008). However, this form of creativity has to be introduced and developed from outside of the game, whereas in *Minecraft* the conditions of creativity are coded into the game itself, in terms of its rules (a distinct creative mode) and mechanics (the world being composed entirely of blocks, which the player can manipulate).

In relation to non-videogame interfaces, creating new horizons of collective secondary protention might involve exposing the artificial, arbitrary nature of the relationship between entities in an interface, and gaining an understanding of how objects are designed to express a particular resolution and technicity. As *Minecraft* exposes the artificiality of its world, it also gives the player a sense of control in that they can understand the principles behind its structure and respond to these principles in their attempts to build what they like. In the same way, non-gaming interfaces such as smart meter apps could be developed that do not simply measure and compare individual energy use but allow users to build and create new ways of presenting information regarding energy. Rather than creating arbitrary distinctions between individual customers as the basic unit for comparison, such devices could show energy consumption for streets or neighbourhoods as a whole. In turn, the presentation of this information could potentially cultivate social relations between members of a community that are based on solidarity and help enable discussion regarding bargaining for collective energy purchasing.

## Pharmacology

Of course, it must be emphasized that all of the techniques and technologies of the envelope that have been described throughout the book can be understood as what Stiegler (2013b), drawing upon the work of Derrida (2004), terms a *pharmakon*. A *pharmakon* refers to an object that is both a poison and cure, depending on how it is used. More specifically, *pharmakon* refers to a continuing movement between 'autonomy and heteronomy' (Stiegler 2013b: 2) in which an object is both autonomous from other things, yet central to the composition of those things. Stiegler discusses the concept of the *pharmakon* in relation to Winnicott's (1953) understanding of transitional objects. He suggests: 'the transitional object is the first *pharmakon* because it is both an external object on which the mother and child are dependent...and in relation to which they are thus heteronomous; and an object that, not existing but consisting, provides (through this very consistence) sovereignty to both mother and child' (Stiegler 2013b: 2–3). As Stiegler argues, all technical objects have this pharmacological status. They both open up horizons of possibility, by shaping the relationship between retentions and protentions and limit and enclose these possibilities according to the logics of those who made, assembled or designed them.

For instance, playing videogames can work as pain distraction and ease the stress of medical and hospital procedures (Patel et al. 2006, Jameson et al. 2011), or they can be used to train surgeons to perform more accurately during surgery (Kato 2010). Interface envelopes can also be productively employed to reshape practices of social care and being together. In their case study of a child patient playing *Minecraft* over a long stay in hospital, Hollett and Ehret (2014: 2) suggest that the game helped cultivate an atmosphere in which 'the social textures of being in the hospital were (re)shaped: Care for Parker [the patient being observed], and by Parker, was amplified'. Here, games, via the interface envelopes they potentially generate 'hold...the potential to alter the felt-relations between bodies, attuning them to each other, and, potentially, evoking a deeper sense of each others' bodies' (Hollett and Ehret 2014: 2). More speculatively, one could imagine the development of programs for augmented reality technology such as Google Glass that generate a very narrow interface envelope that could benefit people living with Alzheimers and Dementia (Vradenburg 2014). Creating an interface envelope that organized perception around a very narrowly spaced set of tertiary retentions, that reminds patients what to do and where to go, might help alleviate the fundamental disorientation that makes living with dementia so difficult.

Living with envelope life means recognizing the pharmacological aspects of digital interfaces and working to retool them to create conditions for the emergence of new modes of nowness or ways of utilizing the nowness of envelopes for helpful purposes. Ironically, the very envelopes that attempt to narrow the now as a mode of experience also provide the perceptual tools to combat this narrowing. Even as envelopes narrow the now, they also stimulate users' capacities to sense the difference between ever more finely distinguished now points. This capacity to sense difference provides the seeds of opportunity to create new economies of protention, whereby the now is opened up rather than closed down, and through which alternative ways of thinking and knowing the future become possible.

# 8

# Conclusions

In this book, I have argued that interfaces are constructed from inorganically organized objects, which are designed to emanate particular resolutions and technicities. The resolutions and technicities that emerge from objects are tied to an object's particular capacities as much as they are constituted relationally, in that they depend upon their encounters with other objects. Games and interface designers organize these objects in such a way as to attempt to create resolutions and technicities that cultivate particular skills, habits, affects, actions and engagement from players. I have termed this organization and distribution of objects an 'interface environment'. Engaging with these environments in turn can create envelopes of spatio-temporal perception. These envelopes can be understood as localized foldings of space and time around which the relationship between users' memory and anticipation is organized for the explicit creation of profit for the industries that produce these interfaces.

As I have suggested, a trend in big-budget contemporary videogame design attempts to organize players' perception around a continuously modulating now point in order to keep them captivated and playing particular games. In contrast to writers, such as Rushkoff (2013), who argue that digital technology encourages a perpetual present tense, I have argued that games designers do not attempt to organize perception around a single moment. Instead, they actively attempt to fold the present into spatio-temporal envelopes of differing lengths. The length of these envelopes differs between games: in *Street Fighter IV*, these envelopes may be half a second, in *Call of Duty*, they may be ten seconds, while in *Final Fantasy XIII*, they may be between fifteen to twenty seconds. Successfully constructed envelopes can overlap, creating experiences that last minutes or hours, during which time the players' spatio-temporal perception is organized around a now where reflecting upon the past or projecting their thought into the future is minimized.

The simple aim of this complex procedure is to create envelope power from the interface. Drawing upon and modifying Neidich and Stiegler's accounts of neuropower and psychopower, envelope power can be defined as an attempt by interface designers and publishers to increase users' capacities to sense difference and use this capacity to alter the relationship between anticipation and memory in order to keep users engaged in the products they sell. In the videogames industry, this envelope power is based around the construction and maintenance of player attention, which becomes the key currency that developers can use to create commercially successful games.

Interface envelopes are not inevitable consequences of playing a game, nor are they 'hard coded' into the game's mechanics or software. Rather, they are an emergent effect that only appears through players' active engagement with an interface. Although designers attempt to shape players' actions through the resolution and technicity of the objects they design and the encounters they prime with these objects, the interactive nature of the interface means that there is always a contingency of player action that designers cannot eliminate. Indeed, this contingency is a large part of the appeal of gaming compared to other types of media, which are determined in advance, such as film. As such, these envelopes do not simply determine what players do or how they feel while playing. But, at the same time, the envelopes generated between player and game are never completely accidental or contingent upon the singularity of a particular player or encounter. Games designers can modulate the resolution and technicity of objects to produce a particular space of possibilities in an attempt to generate the outcomes they desire. The fact that many of the games discussed in this book have been both commercially and critically successful suggests that the designers of these games have been at least partially effective in creating the conditions for the production of interface envelopes and envelope power.

In the rest of this concluding chapter, I want to build upon the points I made in the previous chapters to think through the implications of these arguments for broader debates surrounding three key areas of concern: games and digital interfaces; objects and new materialism; and accounts of cognitive capitalism, power and economy. Reflecting on these three areas allows us to think beyond games design to consider how an account of the interface envelope allows us to understand other services and products that are being created under cognitive capitalism and how we might theorize these objects and services.

## Games/digital interfaces

Reflecting on accounts of space and time within game studies, I have suggested there is no 'space' or 'time' in videogames or any other screen-based environment. There are only processes of spacing or timing, where something such as space or time appears as a particular kind of phenomenon through the construction of relations and non-relations between objects that make up the interface. This is rather different from existing accounts of game space that consider game space to be the outcome of multiple 'types' of space or a cognitive phenomenon, imagined mainly in the mind of the player. For example, Nitchse (2008: 15–16) outlines a five-layer model of game space. Here, the space of games includes the rule-based space of graphics and AI, the mediated space of the image on screen, the fictional space 'imagined by players from their comprehension of available images', the play space (the physical space in which body and console are situated) and the social space (the relationships with other players, such as in multiplayer games).

Developing an account of inorganically organized objects that emanate a resolution and technicity suggests that game worlds or spaces are not imagined or immaterial worlds created in the mind or somewhere beyond the screen. They are better understood as actual worlds, composed of objects that shape the space and time that appear to constitute those worlds as worlds. The technicity that emerges through the distribution of things shapes how the now emerges as a unit of temporality for both the player engaging with the game and the other objects involved in the encounter. For the human player, this technicity also works to organize perception through constructing the relationship between memory and anticipation.

In turn, the resolution of objects shapes how space appears. How objects are distributed and differentiated via their qualities generates the resolution and thus the spatiality of an environment as a whole. The space of a game environment, such as Seattle in *Infamous: Second Son*, is not a container within which the action of the game takes place. Rather, it is a distribution of objects – ground, trees, buildings, cars, rivers and so on – that encounter and transduce qualities to one another, through which the particular capacities of these objects emerge. Walls block access, neon signs can be drained to gain power and Delsin, while the avatar cannot swim in water, can only jump to a certain height and can only run a certain speed. Space appears as the space between these objects, and the player is drawn towards or discouraged away from objects depending on their resolution and the qualities they emanate.

However, this account does not simply reduce videogame space and time to an assemblage of material objects either. As I discussed in Chapter 2, Taylor (2009) draws upon an actor network perspective to suggest that game interfaces emerge from the relationship between a variety of objects, including software, hardware, social relations, imaginings of the player and so on. While in some respects Taylor's description is accurate, the space and time of interface environments are not reducible to assemblages of matter because objects never fully contact or meet one another and the qualities that emerge through transductions between objects are specific to those encounters and transductions. Within any interface environment, there is always the potential for a new encounter between objects to occur that would in turn create a different resolution or technicity and so change the space and time of the interface. In other words, space and time in interfaces are not reducible to the sum total of relations between objects that make up an interface because there is never a sum total of fully disclosed objects in any environment. Or, more specifically, the objects that make up an environment always have the capacity to encounter one another in a different way and thus disclose new qualities or produce a new object, however much designers might attempt to control how objects encounter one another within the interface, according to their own aims and goals.

With this perspective in mind, we can argue that videogame environments (or any screen-based digital interface) are not ontologically different from other kinds of environment or space. What differentiates environments or spaces from one another is how objects and beings can access other objects that make up that environment. In videogames access to the space or environment is constituted by the control pad or interface device, which generally offers more limited capacities for engaging with objects than moving around and manipulating objects in everyday, non-videogame settings.

In relation to work on digital interfaces more broadly, this perspective would indicate the need to avoid making distinctions between actual and virtual or real and imagined space as distinct categories or realms of being. As Kinsley (2013) and Graham (2013) argue, such terms remain popular in accounts of augmented reality and screen and mobile media (see also Richardson 2009, Farman 2010, 2012, Watkins et al. 2012). But, as I suggested in Chapter 2, these distinctions reduce objects in the interface to representations or images that only become meaningful when experienced by a subject. As my analysis of envelopes and envelope power has shown, considering objects in interfaces as real objects opens up productive ways to understand how they shape the neural and habitual capacities of users' bodies that an account based on the 'immateriality' of cyberspace might have more difficulty with. To analyse digital interfaces more broadly, we can begin by investigating the kinds of objects that make up interfaces and the technicities and resolutions these objects emanate and compare these with the objects that make up other kinds of environment.

# CONCLUSIONS

## Objects/new materialism/envelopes

This book has classified and investigated a particular type or category of object, which it has termed, following Stiegler, 'inorganically organised objects'. This account has moved away from a subject- or human-centred account of things, where objects exist only for and through the ways that humans access or use these objects. In this case, objects can't be reduced, as Heidegger (1962) argues, to the ways humans access them as either ready-to-hand or present-at-hand. Objects in interface environments have a capacity to encounter one another and transduce qualities outside of the correlate of intentional or human experience. In many ways, this claim is quite obvious. But, this simplicity is somewhat misleading. By developing the concepts of technicity and resolution, I have complicated this account by showing how the transductions that occur when objects encounter one another also shape the appearance of space and time to humans (and other objects) that engage with these things.

All inorganically organized objects, digital or otherwise, have a technicity and a resolution and these terms are not just specific to digital interfaces. Indeed, thinking about the technicity and resolution of objects beyond interfaces opens new avenues to theorize and study technical things more generally. I am not, then, advocating an account of technology from a phenomenological or actor network theory perspective, in which things are considered either as practical or hybrid tools (Verbeek 2008) or as non-dualistic assemblages of entities (Prout 1996, Murdoch 1997). Instead, I use the concepts of technicity and resolution to attend to what Bennett (2009) calls 'thing power', in which an object's capacity to affect is more than the sum of its relations with other things, even if it remains to some degree relative to human use. Developing this kind of post-phenomenology offers broader opportunities to study human–object and object–object relations. For example, a post-phenomenologist could study car design in order to understand how the resolution and technicity of the steering wheel affects driver safety, or how the resolution and technicity of a firearm affects its user's decision-making process in choosing to fire that weapon. Beginning with the way objects appear to humans and other objects, we can use the concepts of resolution and technicity to speculate about the types of space and time they open up and the forms of memory and anticipation they encourage.

Envelopes in videogames are the particular product of a carefully designed and distributed set of objects that form an environment. Envelopes are possible in games and digital interfaces largely because the player voluntarily enters into a relation with an interface environment, which makes the emergence of an envelope more likely. But, it is important to remember that other non-digital inorganically organized objects, such as a hammer, can also generate envelopes. However, where the interface envelope of the hammer is an inadvertent by-product of using the hammer, the interface envelopes that emerge from digital interfaces are actively encouraged by the designers or creators of those technologies. As I argued in Chapter 5, the director of *Final Fantasy XIII* tried to design the battle system to create an envelope that is fifteen to twenty seconds in length. Using a hammer may create an envelope six seconds in length, but the hammer designer did not have this envelope in mind when making the object. As videogame design becomes more accomplished and the mechanics of game design spread further into everyday life through objects such as wearable technology and smart meters, so the specificity of the envelopes that designers attempt to create may also become more exacting. How effective, durable, repeatable and widespread these forms of envelope power become, however, remains to be seen.

The return to objects initiated by debates in object-orientated ontology and new materialism has been subject to critique. Accounts of objects as discrete units that precede their relations with other things have been questioned by social scientists and theorists, for whom relationality or relationism is the bedrock for thinking social life (Galloway 2013b). The account of inorganically organized objects developed here has attempted to tread the ground between two distinct understandings of objects. The first considers objects to ultimately precede their relations with other objects (Harman 2010a, 2010b, 2011). The second considers objects to be the sum of the parts or relations that constitute them (Anderson and Wylie 2009). Developing an account of technicity and resolution, I have suggested that while objects have a technicity and resolution that precede their encounter with other things, the resolution or technicity that emerges in a given encounter is also specific to that encounter. The qualities of an object that are transduced in an encounter are both shaped by its capacities and its encounters. Therefore, the object can be reduced to neither its capacities nor its encounters alone. From a flat ontological perspective, this means taking all objects in a situation equally seriously, while recognizing that these objects can be differentiated by their qualities and encounters and are therefore not practically equal to one another.

Galloway argues that considering objects from an ontological perspective, as units that precede their relations with other objects, risks defending a philosophical absolute that ignores material history. He asks the following questions that he sees as core to the debate between relational and non-relational accounts of objects:

> Is a philosopher following an ontological absolute or following material history? Do real networks of object relations produce history, or does history produce real networks of objects relations? The answer to the question will indicate how any given person stands in today's debate. Either one prizes pure ontology in the form of the absolute, the One, the infinite, what one used to call God. Or one prizes the historicity of worlds, saturated as they are with all the details of material life. In short, the 'real' in philosophical realism means the absolute. Whereas for a materialist, the 'real' means history. (Galloway 2013: 364)

While Galloway (2013b: 364) accepts that this framing 'invites a misunderstanding', he is keen to emphasize that 'the issue is not objects versus humans but rather the real versus history. For, as we know, objects have histories just like we do'. Galloway (2013b: 366) suggests that choosing a side in this distinction is a political move, and that a non-relational realist account of objects is ultimately apolitical because it focuses on 'abstraction, logical necessity, universality, essence, pure form, spirit, or idea' over rooting everything in 'material life and history'.

The account of inorganically organized objects and the way they are designed to emanate resolutions and technicities I have developed in this book is an attempt to avoid such formal distinctions between realism and abstraction on the one hand and materialism and history on the other. The objects that make up digital interfaces are made by humans, whether they are hardware manufactured in factories or digital weapons modelled in software systems by games designers. To be sure, the processes and logics that went into them have a history that can be traced and mapped in relation to specific economic and ideological forces, groups and institutions. But, at the same time, inorganically organized objects have a capacity to act that exceeds the humans that manufactured, modelled and designed them.

Indeed, the whole difficulty of designing digital game interfaces is attempting to anticipate what these objects will do and how they will be used once they have left the factory or been printed onto discs and are being played 'in the wild' (Hutchins 1996, Ash 2010a). Understanding these capacities does require some form of speculative theoretical abstraction, otherwise the capacity for game objects to exist or act is diminished to a humanist instrumentalism in which the object is reduced to either the intentions of the designers that made them or the experience a player has when encountering them. As I have argued across the previous chapters, this is not true in relation to how players respond to interfaces. Objects regularly confound players, designers and software and hardware alike and cannot be reduced to a sum or product of player, designer or hardware, nor to the relations between them.

Quoting Power (2009: n.p.), Galloway asks 'what use is...[realism]...if it simply becomes a race to the bottom to prove that every entity is as meaningless as every other'. However, developing a flat ontology in which inorganically organized objects have a capacity to transduce qualities outside of the way they appear to humans or are created by designers does not mean that every object is equal or meaningless. While a flat ontology might suggest that humans and objects exist on the same ontological level, it does not mean that humans and objects have the same capacity to affect and be affected or encounter one another. Rather, understanding objects through what I have termed a 'post-phenomenological perspective' allows us to take non-human objects seriously, without discarding a notion of the human as having distinctive capacities from other types of being.

## Power/economy/capitalism

The concepts of resolution, technicity, envelopes and envelope power developed in this book have introduced theoretical and empirical distinctions that complicate academic accounts of neuropower, psychopower, affective labour and cognitive capitalism. Both Neidich (2010, 2013) and Stiegler (2010a) develop concepts to analyse how power works to modulate the habitual, neural and conscious capacities of the body and brain to capture and hold attention and create captivated consumers. I have suggested that, while helpful in various ways, Neidich and Stiegler are ultimately too pessimistic regarding the actual effects neuropower and psychopower have on the bodies and brains of people using these technologies. Furthermore, both Neidich and Stiegler's accounts are based upon a normative understanding of a neuro-typical human body that marginalizes and perhaps even dehumanizes those who do not display these capacities. In their pessimism, Neidich and Stiegler then miss the more productive ways that technologies operate to increase users' pre-existing capacities and how these technologies actively cultivate new capacities to act as well. Envelope power differs from either neuropower or psychopower in that it does not simply attempt to capture a pre-existing consciousness, but actively creates conditions for the production of new forms of consciousness that can differentiate between increasingly small units of space and time. Through creating these forms of consciousness, the interface industries can then use this micro-differentiation to create more profit from the products they create.

Questioning or critiquing homeomorphic envelope power is not a matter of attempting to disentangle ourselves from it. Rather, drawing upon Stiegler's concept of the *pharmakon*, we can recognize that, in Van Camp's (2012: n.p.)

words: 'the only way to confront contemporary...power is by re-inventing this *same* mnemotechnical system...[that creates envelope power]...in such a way that it enables the emergence of a new culture of care. Any critical response to the current mnemotechnical system must arise from within its own possibilities'. As I examined in Chapter 7, questioning the logics inherent to envelope power does not have to involve the user trying to avoid engaging with the technologies of the interface envelope. Instead, players can look to alternative ways of being-in-envelopes that cultivate social relations with others and encourage the development of secondary retentions that do not conform to the intentions or designs of the interface industries.

In relation to affective labour, this book has argued that capital attempts to increase and create economic value in its products through encouraging and stimulating users' capacities to sense difference. In the videogames I have examined, this capacity to sense difference is organized around the ability of players to distinguish between increasingly small units of space and time. Furthermore, through statistical and tracking systems, I also argued that videogames attempt to defer and suspend the negative affects of players in order to keep them playing the game. Taken together, we can consider how affect is utilized by capital in order to increase the perceived value of the products companies create. Increasing value is not simply about increasing players' positive affects, by making them feel happy or joyful all of the time. Rather, it is about creating envelopes that attempt to manipulate players' affects within a set of possibilities that the designers of these technologies have identified as most useful or productive for the creation of emotionally attached and attentive consumers. As Boutang (2012) argues in *Cognitive Capitalism*, the latest stage of capitalist development is not concerned with the production of physical goods and commodities, but about capturing value from affective and cognitive attention. As this book has attempted to show, this creation of value does not simply arise from the capture of pre-existent attention but creates new capacities for attention as well. In dividing space and time into ever smaller units of temporality and spatiality, the capitalist system also creates new territories of attention to be mined and exploited.

Within videogames, the development of specific sensory capacities could be linked to broader debates around the rise of an individualized neoliberal subject (Katz 2005, Read 2009, Gane 2014), whose existence is brought into being by these technologies (amongst others). Indeed, online systems such as *Call of Duty Elite* or *Halo Waypoint* seem to point to how, even in multiplayer games and settings, players' attention becomes purely focused on their individual attainment irrespective of or in spite of their teammates or those with whom they are playing. It may be possible to suggest that a logic of the interface envelope seeks to create individualized subjects, whose mode of existence is fundamentally shaped by the possibilities afforded to them by the envelopes they inhabit. Aspects of games design are being applied through processes of gamification to an increasing number of work and recreational-based activities. Although currently the outcome of particularly intense game experiences, one could argue that the envelope, and the modes of spatio-temporal perception it encourages, could become a more widespread, banal and diffuse phenomenon that shapes practices and relations in everyday life outside of games. The extent to which this might occur is yet to be seen, but this book has provided a conceptual vocabulary and mode of analysis to begin to think through these processes both within and outside of game interfaces.

# Bibliography

Aarseth, E. J. (1997), *Cybertext: Perspectives on Ergodic Literature*, Baltimore, MD: Johns Hopkins University Press.
Adams, F. M. and C. E. Osgood (1973), 'A Cross-Cultural Study of the Affective Meanings of Color', *Journal of Cross-Cultural Psychology*, 4: 135–156.
Allen, J. (2011), 'Topological Twists: Power's Shifting Geographies', *Dialogues in Human Geography*, 1: 283–298.
Amoore, L. (2011), 'Data Derivatives: On the Emergence of a Security Risk Calculus for Our Times', *Theory, Culture and Society*, 28: 24–43.
———. (2013), *The Politics of Possibility*, Durham: Duke University Press.
Anderson, B. (2006), 'Becoming and Being Hopeful: Towards a Theory of Affect', *Environment and Planning D: Society and Space*, 24: 733–752.
———. (2012), 'Affect and Biopower: Towards a Politics of Life', *Transactions of the Institute of British Geographers*, 37: 28–43.
Anderson, B. and P. Harrison (2012), *Taking-place: Non-Representational Theories and Geography*, Farnham: Ashgate Publishing, Ltd.
Anderson, B. and J. Wylie (2009), 'On Geography and Materiality', *Environment and Planning. A*, 41: 318.
Antic, D. and M. Fuller (2011), 'The Computation of Space', C. U. Andersen and S. B. Pold (eds), *Interface Criticism: Aesthetics Beyond the Buttons*, Aarhus: Aarhus University Press.
Apperley, T. and D. Jayemane (2012), 'Game Studies' Material Turn', *Westminster Papers in Communication and Culture*, 9: 5–26.
Ash, J. (2009), 'Emerging Spatialities of the Screen: Video Games and the Reconfiguration of Spatial Awareness', *Environment and Planning A*, 41: 2105–2124.
———. (2010a), 'Architectures of Affect: Anticipating and Manipulating the Event in Processes of Videogame Design and Testing', *Environment and Planning D: Society and Space*, 28: 653–671.
———. (2010b), 'Teleplastic Technologies: Charting Practices of Orientation and Navigation in Videogaming', *Transactions of the Institute of British Geographers*, 35: 414–430.
———. (2012a), 'Attention, Videogames and the Retentional Economies of Affective Amplification', *Theory, Culture and Society*, 29: 3–26.
———. (2012b), 'Technology, Technicity, and Emerging Practices of Temporal Sensitivity in Videogames', *Environment and Planning A*, 44: 187–204.
———. (2013a), 'Rethinking Affective Atmospheres: Technology, Perturbation and Space Times of the Non-human', *Geoforum*, 49: 20–28.
———. (2013b), 'Technologies of Captivation Videogames and the Attunement of Affect', *Body and Society*, 19: 27–51.

# BIBLIOGRAPHY

———. (2014), 'Technology and Affect: Towards of Theory of Inorganically Organised Objects', *Emotion, Space and Society*. doi: 10.1016/j.emospa.2013.12.017.

Ash, J. and P. Simpson (2014), 'Geography and Post-phenomenology', *Progress in Human Geography*. doi: 10.1177/0309132514544806.

Ashcraft, B. (2010), 'Let The Beast Show You How to Hold Your Stick'. Retrieved 21 July 2014, from http://kotaku.com/5626160/let-the-beast-show-you-how-to-hold-your-stick

Barad, K. (2007), *Meeting the Universe Halfway: Quantum Physics and the Entanglement of Matter and Meaning*, Durham: Duke University Press.

Barker, M. and J. Petley (2002), *Ill Effects: The Media Violence Debate*, New York, NY: Routledge.

Becker, K. (2009), 'The Power of Classification: Culture, Context, Command, Control, Communications, Computing', K. Becker and F. Stalder (eds), *Deep Search: The Politics of Search Beyond Google*, Munich: Studienverlag & Transaction.

Behrenshausen, B. G. (2007), 'Toward a (Kin)Aesthetic of Video Gaming: The Case of Dance Dance Revolution', *Games and Culture*, 2: 335–354.

Beller, J. (2006), *The Cinematic Mode of Production: Attention Economy and the Society of the Spectacle*, Lebanon: Dartmouth College Press.

Bennett, J. (2009), *Vibrant Matter: A Political Ecology of Things*, Durham: Duke University Press.

Berry, D. M. (2011), *The Philosophy of Software: Code and Mediation in the Digital Age*, London: Palgrave.

Blackman, L. (2013), 'Habit and Affect: Revitalizing a Forgotten History', *Body and Society*, 19: 186–216.

Bogost, I. (2007), *Persuasive Games: The Expressive Power of Videogames*, Cambridge, MA: MIT Press.

———. (2008), *Unit Operations: An Approach to Videogame Criticism*, Cambridge, MA: MIT Press.

———. (2011a), 'Gamification Is Bullshit'. Retrieved 9 July 2014, from http://bogost.com/writing/blog/gamification_is_bullshit/

———. (2011b), *How to Do Things with Videogames*, Minneapolis: University of Minnesota Press.

———. (2012), *Alien Phenomenology, or, What It's Like to Be a Thing*, Minneapolis: University of Minnesota Press.

Bolter, J. (2007), 'Digital Essentialism and the Mediation of the Real', H. Philipsen and L. Qvortrup (eds), *Moving Media Studies: Remediation Revisited*, Frederiksberg: Samfundslitteratur Press: 195–210.

Borgmann, A. (2009), *Technology and the Character of Contemporary Life: A Philosophical Inquiry*, Chicago, IL: University of Chicago Press.

Boutang, Y. M. (2012), *Cognitive Capitalism*, London: Polity Press.

Bradley, A. and L. Armand (2006), *Technicity*, Prague: Litteraria Pragensia.

Brophy, E. (2011), 'Language Put to Work: Cognitive Capitalism, Call Center Labor, and Worker Inquiry', *Journal of Communication Inquiry*, doi: 10.1177/0196859911417437.

Bruce, J. W. and P. J. Giblin (1984), *Curves and Singularities*, Cambridge: Cambridge University Press.

Bryant, J. and M. Oliver (2009), *Media Effects: Advances in Theory and Research*, New York, NY: Routledge.

Bryant, L. R. (2011), *The Democracy of Objects*, Chicago, IL: MPublishing.

Bryant, L., N. Srnicek and G. Harman (2011), *The Speculative Turn: Continental Materialism and Realism*, Chicago, IL: re.Press.

Carter-White, R. (2009), 'Auschwitz, Ethics, and Testimony: Exposure to the Disaster', *Environment and Planning D: Society and Space*, 27: 682–699.

Chapple, C. (2014), 'Can the Game Industry Keep a Lid on Rising Development Costs?'. Retrieved 9 June 2014, from http://www.develop-online.net/analysis/can-the-game-industry-keep-a-lid-on-rising-development-costs/0192815

Cheney-Lippold, J. (2011), 'A New Algorithmic Identity: Soft Biopolitics and the Modulation of Control', *Theory, Culture and Society*, 28: 164–181.

Chesher, C. (2012), 'Navigating Sociotechnical Spaces: Comparing Computer Games and Sat Navs as Digital Spatial Media', *Convergence: The International Journal of Research into New Media Technologies*, 18: 315–330.

Chesney, T., S. -H. Chuah, R. Hoffmann, W. Hui and J. Larner (2014), 'Skilled Players Cooperate Less in Multi-player Games', *Journal of Gaming and Virtual Worlds*, 6: 21–31.

Chun, W. (2006), *Control and Freedom: Power and Paranoia in the Age of Fiber Optics*, Massachusetts: MIT Press.

Chun, W. H. K. (2011), *Programmed Visions: Software and Memory*, Cambridge, MA: MIT Press.

Clarke, D. (2011), 'Music, Phenomenology, Time Consciousness: Mediations After Husserl', D. Clarke and E. Clarke (eds), *Music and Consciousness: Philosophical, Psychological, and Cultural Perspectives*, Oxford: OUP Oxford: 1–29.

Clough, P. T. (2013), 'The Digital, Labor, and Measure Beyond Biopolitics', T. Scholz (ed), *Digital Labor: The Internet as Playground and Factory*, London: Routledge: 112–126.

Clough, P. T., G. Goldberg, R. Schiff, A. Weeks and C. Willse (2007), 'Notes Towards a Theory of Affect-Itself', *Ephemera*, 7: 60–77.

Cockshut, L. (2012), *The Way We Play: Exploring the Specifics of Formation, Action and Competition in Digital Gameplay Among World of Warcraft Raiders*, PhD, Durham University.

Colebrook, C. (2014), *Death of the PostHuman, Essays on Extinction*, Vol 1, Chicago, IL: Open Humanities Press.

Combes, M. and T. LaMarre (2012), *Gilbert Simondon and the Philosophy of the Transindividual*, Cambridge, MA: University Press Group Limited.

Cramer, F. and M. Fuller (2008), 'Interface', M. Fuller (ed), *Software Studies: A Lexicon*, Cambridge, MA: MIT Press: 149–153.

Crick, T. (2011), 'The Game Body: Toward a Phenomenology of Contemporary Video Gaming', *Games and Culture*, 6: 259–269.

Crogan, P. (2010), 'Bernard Stiegler: Philosophy, Technics and Activism', *Cultural Politics*, 6: 133–156.

———. (2011), *Gameplay Mode: War, Simulation, and Technoculture*, Minneapolis: University of Minnesota Press.

Crogan, P. and H. Kennedy (2009), 'Technologies Between Games and Culture', *Games and Culture*, 4: 107–114.

Crogan, P. and S. Kinsley (2012), 'Paying Attention: Towards a Critique of the Attention Economy', *Culture Machine*, 13: 1–29.
Csikszentmihalyi, M. (2009), *The Evolving Self: Psychology for the Third Millennium, A*, New York, NY: HarperCollins.
Curran, S. (2004), *Game Plan: Great Designs that Changed the Face of Computer Gaming*, Mies: RotoVision.
De Boever, A., A. Murray, J. Roffe and A. Woodward (2012), *Gilbert Simondon: Being and Technology*, Edinburgh: Edinburgh University Press.
Deleuze, G. (1988), *Spinoza: Practical Philosophy*, San Francisco, CA: City Lights Publishers.
———. (1992), 'Postscript on the Societies of Control', *October* 59: 3–7.
———. (2005), *Francis Bacon: The Logic of Sensation*, New York, NY: Continuum.
Depraz, N. (2004), 'Where is the Phenomenology of Attention that Husserl Intended to Perform? A Transcendental Pragmatic-orientated Description of Attention', *Continental Philosophy Review*, 37: 5–20.
Der Derian, J. (2009), *Virtuous War: Mapping the Military-Industrial-Media-Entertainment-Network*, London: Taylor & Francis.
Derrida, J. (1998), *Of Grammatology*, Baltimore, MD: Johns Hopkins University Press.
———. (2004), *Dissemination*, New York, NY: Bloomsbury Academic.
Derrida, J. and B. Stiegler (2002), *Echographies of Television: Flimed Interviews*, Cambridge: Polity.
Deterding, S., R. Khaled, L. Nacke and D. Dixon (2011). *Gamification: Toward a Definition*. CHI, Gamification Workshop Proceedings, 2011. Vancouver, BC, Canada: 7–12.
Dewsbury, J. (2009), *Performative, Non-Representational, and Affect-Based Research: Seven Injunctions. The SAGE Handbook of Qualitative Geography*, London: SAGE Publications Ltd.
Dodge, M. and R. Kitchin (2005), 'Code and the Transduction of Space', *Annals of the Association of American Geographers*, 95: 162–180.
Dodgshon, R. A. (2008), 'In What Way Is the World Really Flat? Debates Over Geographies of the Moment', *Environment and Planning D: Society and Space*, 26: 300–314.
Donovan, T. (2010), *Replay: The History of Video Games*, Lewes: Yellow Ant Media.
Drews, G. (1973), 'Fine Structure and Chemical Composition of Cell Envelopes', N. G. Carr and B. A. Whitton (eds), *The Biology of Blue-green Algae*, California: Univeristy of California Press: 99–116.
Dunagan, J. (2010), 'Politics for the Neurocentric Age', *Journal of Future Studies*, 15: 51–70.
Elwell, J. S. (2014), 'The Transmediated Self: Life Between the Digital and the Analog', *Convergence: The International Journal of Research into New Media Technologies*, 20: 233–249.
Ernst, W. (2013), *Digital Memory and the Archive*, Minneapolis: University of Minnesota Press.
Farman, J. (2010), 'Hypermediating the Game Interface: The Alienation Effect in Violent Videogames and the Problem of Serious Play', *Communication Quarterly*, 58: 96–109.

———. (2012), *Mobile Interface Theory: Embodied Space and Locative Media*, London: Routledge Publishing.

Feffer, M. (2011), 'Gamification in the Real World: VW's Experiment with Fun'. Retrieved 12 June 2014, from http://news.dice.com/2011/12/21/fun-theory/

Foucault, M. (1977), *Discipline and Punish*, London: Penguin Books.

———. (2003), *"Society Must Be Defended": Lectures at the Collège de France, 1975–1976*, London: Penguin.

Frith, J. (2013), 'Turning Life into a Game: Foursquare, Gamification, and Personal Mobility', *Mobile Media and Communication*, 1: 248–262.

Fuchs, M., S. Fizek, P. Ruffino and N. Schrape, eds (2014), *Rethinking Gamification*, Leuphana: Meson Press.

Gallacher, L. A. (2011), '(Fullmetal) Alchemy: The Monstrosity of Reading Words and Pictures in Shonen Manga', *Cultural Geographies*, 18: 457–473.

Galloway, A. R. (2009), 'The Unworkable Interface', *New Literary History*, 39: 931–955.

———. (2013a), *The Interface Effect*, New York, NY: Wiley.

———. (2013b), 'The Poverty of Philosophy: Realism and Post-Fordism', *Critical Inquiry* 39: 347–366.

GameSpot. (2010), 'Final Fantasy XIII Battle System Interview by GameSpot'. Retrieved 11 April 2014, from https://www.youtube.com/watch?v=PVQtiLBhHM0

Gane, N. (2014), 'The Emergence of Neoliberalism: Thinking Through and Beyond Michel Foucault's Lectures on Biopolitics', *Theory, Culture and Society*, 31: 3–27.

Gibson, J. J. (1979), *The Theory of Affordances*, Hilldale, USA: Lawrence Erlbaum Associates, Inc., Publishers.

Giddings, S. (2007), 'Playing with Nonhumans: Digital Games as Technocultural Form', S. de Castells and J. Jensen (eds), *Worlds in Play: International Perspectives on digital Games Research*, London: Peter Lang.

———. (2014), *Gameworlds: Virtual Media and Children's Everyday Play*, New York, NY: Bloomsbury.

Gill, R. (2008), 'Culture and Subjectivity in Neoliberal and Postfeminist Times', *Subjectivity*, 25: 432–445.

Glennie, P. D. and N. Thrift (2002), 'The Spaces of Clock Time', P. Joyce (ed), *The Social in Question: New Directions in History and the Social Sciences*, London: Routledge: 151–174.

Goldhaber, M. H. (1997), 'The Attention Economy and the Net', *First Monday*. 2: no pagination. http://firstmonday.org/article/view/519/440

Graham, M. (2013), 'Geography/Internet: Ethereal Alternate Dimensions of Cyberspace or Grounded Augmented Realities?', *The Geographical Journal*, 179: 177–182.

Gratton, P. (2014), *Speculative Realism: Problems and Prospects*, New York, NY: Bloomsbury.

Gregg, M. and G. J. Seigworth (2010), *The Affect Theory Reader*, Durham: Duke University Press.

Grossman, R. P. and J. Z. Wisenblit (1999), 'What We Know About Consumers' Color Choices', *Journal of Marketing Practice: Applied Marketing Science*, 5: 78–88.

# BIBLIOGRAPHY

Grover, R. and M. Nayak. (2014), 'Activision Plans $500 million Date with "Destiny"'. Retrieved 6 June 2014, from http://www.reuters.com/article/2014/05/06/us-activision-destiny-idUSBREA4501F20140506

Hall, S. (1993), 'Encoding, Decoding', *The Cultural Studies Reader*, 4: 90–103.

Halter, E. (2006), *From Sun Tzu to XBox: War and Video Games*, New York, NY: Thunder's Mouth Press.

Hansen, M. (2000), *Embodying Technesis: Technology Beyond Writing*, Chicago, IL: University of Michigan Press.

———. (2004), *New Philosophy for New Media*, Cambridge, MA: MIT Press.

———. (2006a), *Bodies in Code: Interfaces with Digital Media*, New York, NY: Routledge.

———. (2006b), 'Media Theory', *Theory, Culture and Society*, 23: 297–306.

Hardt, M. and A. Negri (2000), *Empire*, Cambridge: Harvard University Press.

Harman, G. (2002), *Tool-Being: Heidegger and the Metaphysics of Objects*, Chicago, IL: Open Court.

———. (2005), *Guerrilla Metaphysics: Phenomenology and the Carpentry of Things*, Chicago, IL: Open Court.

———. (2009), *Prince of Networks: Bruno Latour and Metaphysics*, Melbourne: re.Press.

———. (2010a), *Circus Philosophicus*, Winchester: Zero Books.

———. (2010b), *Towards Speculative Realism: Essays and Lectures*, Winchester: Zero Books.

———. (2011), *The Quadruple Object*, Winchester: Zero Books.

——— (2012), On Interface: Nancy's Weights and Masses, in *Jean-Luc Nancy and Plural Thinking: Expositions of World, Ontology, Politics and Sense*, Gratton, P. and Moris, M. E. (ed), Albany, NY: SUNY Press.

———. (2013), *Bells and Whistles: More Speculative Realism*, Winchester: Zero Books.

Harrison, P. (2000), 'Making Sense: Embodiment and the Sensibilities of the Everyday', *Environment and Planning D: Society and Space*, 18: 497–517.

Harvey, D. (2005), *A Brief History of Neoliberalism*, Oxford: Oxford University Press.

Hayles, N. K. (2007), 'Hyper and Deep Attention: The Generational Divide in Cognitive Modes', *Profession 2007*, 187–199.

———. (2012), *How We Think: Digital Media and Contemporary Technogenesis*, Chicago. IL: University of Chicago Press.

Heidegger, M. (1962), *Being and Time*, Oxford: Blackwell Publishing.

———. (1982), *The Question Concerning Technology, and Other Essays*, New York, NY: HarperCollins.

———. (1992), *The Concept of Time*, Oxford: Blackwell Publishing.

———. (2006), *Mindfulness*, New York, NY: Bloomsbury Academic.

———. (2012), *Contributions to Philosophy (of the Event)*, Indiana: Indiana University Press.

Heijden, J. (2010), 'Successful Playtesting in Swords and Soldiers'. Retrieved 23 September 2014 from http://www.gamasutra.com/view/feature/5939/successful_playtesting_in_swords__.php

Hillis, K. (2009), *Online a Lot of the Time: Ritual, Fetish, Sign*, Durham: Duke University Press.

Hollett, T. and C. Ehret (2014), '"Bean's World": (Mine) Crafting Affective Atmospheres of Gameplay, Learning, and Care in a Children's Hospital', *New Media and Society*, doi: 10.1177/1461444814535192.

Huntemann, N. and T. Payne, eds (2010), *Joystick Soldiers: The Politics of Play in Military Video Games*, New York, NY: Routledge.

Husserl, E. (1991), *On the Phenomenology of the Consciousness of Internal Time (1893–1917)*, Dordrecht: Kluwer Academic.

Hutchins, E. (1996), *Cognition in the Wild*, Cambridge: MIT Press.

Hutnyk, J. (2012), 'Proletarianisation', *New Formations*, 77: 127–149.

Ihde, D. (2008), 'Introduction: Postphenomenological Research', *Human Studies*, 31: 1–9.

———. (2010), *Heidegger's Technologies: Postphenomenological Perspectives*, New York, NY: Fordham Univ Press.

Introna, L. D. (2011), 'The Enframing of Code: Agency, Originality and the Plagiarist', *Theory, Culture and Society*, 28: 113–141.

Ip, B. (2008), 'Technological, Content, and Market Convergence in the Games Industry', *Games and Culture*, 3: 199–224.

Isin, E. F. (2004), 'The Neurotic Citizen', *Citizenship Studies*, 8: 217–235.

James, W. (2011), *The Principles of Psychology*, New York, NY: Cosimo Books.

Jameson, E., J. Trevena and N. Swain (2011), 'Electronic Gaming as Pain Distraction', *Pain Research and Management* 16: 27–32.

Jay, M. (1993), *Downcast Eyes: The Denigration of Vision in Twentieth-Centruty French Thought*, Berkeley: University of California Press.

Jeannerod, M. (2005), *Le Cerveau Intime*, Paris: Odile Jacob.

Jenkins, H. (2008), *Convergence Culture*, New York, NY: New York University Press.

Jones, S. and G. Thiruvathukal (2012), *Codename Revolution: The Nintendo Wii Platform*, Cambridge, MA: MIT Press.

Jørgensenn, K. (2013), *Gameworld Interfaces*, Cambridge, MA: MIT Press.

Juul, J. (2005), *Half-real: Video Games between Real Rules and Fictional Worlds*, Cambridge, MA: MIT Press.

Karatzogianni, A. and A. Junstman (2012), *Digital Cultures and the Politics of Emotion: Feelings, Affect and Technological Change*, London: Sage Publishing.

Kato, P. M. (2010), 'Video Games in Health Care: Closing the Gap', *Review of General Psychology*, 14: 113.

Katz, C. (2005), 'Partners in Crime? Neoliberalism and the Production of New Political Subjectivities', *Antipode*, 37: 623–631.

Katz, J. (1999), *How Emotions Work*, Chicago, IL: University of Chicago Press.

Katz, J. (2013), 'Mobile Gazing Two-ways: Visual Layering as an Emerging Mobile Communication Service', *Mobile Media and Communication*, 1: 129–133.

Keogh, B. (2014), 'Across Worlds and Bodies: Criticism in the Age of Video Games', *Journal of Games Criticism*, 1: 1–26.

King, R. (2013), 'COD: Ghosts – Mark Rubin On Fighting Lag, Series Perception and Dog Memes'. Retrieved 21 July 2014, from http://www.nowgamer.com/features/2148027/cod_ghosts_mark_rubin_on_fighting_lag_series_perception_and_dog_memes.html

Kinsley, S. (2010), Representing 'Things to Come': Feeling the Visions of Future Technologies. *Environment and Planning A*, 42(11): 2771–2790.

———. (2013), 'The Matter of "Virtual" Geographies', *Progress in Human Geography*, 38: 364–384.
Kirkpatrick, G. (2009), 'Controller, Hand, Screen: Aesthetic Form in the Computer Game', *Games and Culture*, 4: 127–143.
———. (2013), *Computer Games and the Social Imaginary*, London: Polity.
Kirschenbaum, M. G. (2008), *Mechanisms: New Media and the Forensic Imagination*, Cambridge, MA: MIT Press.
Kitchin, R. and M. Dodge (2011), *Code/Space: Software and Everyday Life*, Cambridge, MA: MIT Press.
Kittler, F. (2009), 'Towards an Ontology of Media', *Theory, Culture and Society*, 26: 23–31.
Koster, R. (2005), *A Theory of Fun for Games Design*, Arizona: Paraglyph Press.
Kow, Y. and B. Nardi (2010), 'Who Owns the Mods?', *First Monday*, 15: no pagination.
Krummel, J. (2006), 'Spatiality in the Later Heidegger: Turning – Clearing – Letting', *Existentia: An International Journal of Philosophy*, XVI: 405–424.
Kucklich, J. (2005), 'Precarious Playbour: Modders and the Digitial Games Industry', *Fibreculture Journal*, 5: no pagination.
———. (2009), 'Virtual Worlds and Their Discontents: Precarious Sovreignity, Governmentality and the Ideology of Play', *Games and Culture*, 4: 340–352.
Lammes, S. (2008), 'Spatial Regimes of the Digital Playground: Cultural Functions of Spatial Practices in Computer Games', *Space and Culture*, 11: 260–272.
Lapworth, A. (2013), 'Habit, Art, and the Plasticity of the Subject: The Ontogenetic Shock of the Bioart Encounter', *Cultural Geographies*, doi:10.1177/1474474013491926.
Lash, S. and C. Lury (2007), *Global Culture Industry: The Mediation of Things*, New York, NY: Wiley.
Latour, B. (1988), *The Pasteurization of France*, Cambridge: Harvard University Press.
———. (2011), 'A Cautious Prometheus? A Few Steps Towards a Philosophy of Design with Special Attention to Peter Sloterdijk', W. Schinkel and L. Noordegraaf-Eelens (eds), *In Medias Res: Peter Sloterdijk's Spherological Poetics of Being*, Amsterdam: Amsterdam University Press: 151–165.
Law, J. (2002), 'Objects and Spaces', *Theory, Culture and Society*, 19: 91–105.
Lazzarato, M. (2006), 'The Concepts of Life and the Living in the Societies of Control', M. Lazzarato (ed), *Deleuze and the Social*, Edinburgh: Edinburgh University Press: 171–190.
Leach Waters, C. (2005), 'The United States Launch of the Sony PlayStation2', *Journal of Business Research*, 58: 995–998.
LeJacq, Y. (2014), 'Here's the Next-Gen Gaming Toothbrush'. Retrieved 12 June 2014, from http://kotaku.com/heres-the-next-gen-gaming-toothbrush-1557707402
Lemke, T. (2014), 'New Materialisms: Foucault and the "Government of Things"', *Theory, Culture and Society*, doi: 10.1177/0263276413519340.
Lowood, H. (2006), 'High-performance Play: The Making of Machinima', *Journal of Media Practice*, 7: 25.
———. (2008), 'Found Technology: Players as Innovators in the Making of Machinima', T. Macpherson (ed), *Digital Youth, Innovation and the Unexpected*, Cambridge, MA: MIT Press.

Lunenfeld, P. (2000), *The Digital Dialectic: New Essays on New Media*, Cambridge, MA: MIT Press.
Mackenzie, A. (2002), *Transductions: Bodies and Machines at Speed*, London: Continuum.
Malabou, C. (2008), 'Addiction and Grace: Preface to Félix Ravaisson's *of Habit*', F. Ravaisson (ed), *Of Habit*, New York, NY: Continuum.
———. (2009), *What Should We Do with Our Brain?*, New York, NY: Fordham University Press.
———. (2012), *The New Wounded: From Neurosis to Brain Damage*, New York, NY: Fordham University Press.
———. (2013), *Plasticity at the Dusk of Writing: Dialectic, Destruction, Deconstruction*, New York, NY: Columbia University Press.
Malaby, T. (2006), 'Parlaying Value: Capital in and Beyond Virtual Worlds', *Games and Culture*, 1: 141–162.
Manovich, L. (2001), *The Language of New Media*, Cambridge, MA: MIT Press.
———. (2013), *Software Takes Command*, New York, NY: Bloomsbury.
Massumi, B. (2002), *Parables for the Virtual: Movement, Affect, Sensation*, Durham: Duke University Press.
McCormack, D. P. (2003), 'An Event of Geographical Ethics in Spaces of Affect', *Transactions of the Institute of British Geographers*, 28: 488–507.
———. (2014), 'Atmospheric Things and Circumstantial Excursions', *Cultural Geographies*, doi: 10.1177/1474474014522930.
Meillassoux, Q. (2010), *After Finitude: An Essay on the Necessity of Contingency*, New York, NY: Bloomsbury Academic.
Merleau-Ponty, M. (2002), *Phenomenology of Perception*, London: Taylor & Francis.
Microsoft. (2011), 'Microsoft Kinect'. Retrieved 5 February 2012, from http://www.youtube.com/watch?v=f5Ep3oqicVU
Millington, B. (2009), 'Wii Has Never Been Modern: "Active" Videogames and the "Conduct of Conduct"', *New Media and Society*, 11: 621–640.
Misazam. (2010), '"Battlefield: Bad Company 2" – Exclusive Interview with Audio Director Stefan Strandberg'. Retrieved 24 April 2014, from http://designingsound.org/2010/03/battlefield-bad-company-2-exclusive-interview-with-audio-director-stefan-strandberg/
Montfort, N. and I. Bogost (2009), *Racing the Beam: The Atari Video Computer System*, Cambridge, MA: MIT Press.
Moores, S. (2012), *Media, Place and Mobility*, London: Palgrave Macmillan.
Morley, D. (1993), 'Active Audience Theory: Pendulums and Pitfalls', *Journal of Communication*, 43: 13–19.
Morton, T. (2013), *Realist Magic: Objects, Ontology, Causality*, Ann Arbor, MI: University of Michigan Library: Open Humanities Press.
Munster, A. (2006), *Materializing New Media: Embodiment in Information Aesthetics*, Lebanon: Dartmouth College Press.
———. (2011), 'From a Biopolitical "Will to Life" to a Noopolitical Ethos of Death in the Aesthetics of Digital Code', *Theory, Culture and Society*, 28: 67–90.
Murdoch, J. (1997), 'Inhuman/Nonhuman/Human: Actor-network Theory and the Prospects for a Nondualistic and Symmetrical Perspective on Nature and Society', *Environment and Planning D*, 15: 731–756.

Murphie, A. and J. Potts (2003), *Culture and Technology*, London: Palgrave Macmillan.
Nancy, J. L. (2000), *Being Singular Plural*, California: Stanford University Press.
———. (2007), *Listening*, New York, NY: Fordham University Press.
———. (2009), *Corpus*, New York, NY: Fordham University Press.
Neidich, W. (2002), *Blow-Up: Photography and the Brain*, New York, NY: Distributed Art Publishers.
———. (2010), 'From Noopower to Neuropower: How Mind Becomes Matter', D. Hauptmann and W. Neidich (eds), *Cognitive Architecture: From Bio-politics to Noo-Politics*, Rotterdam: 010 Publishers.
———. (2013), 'Neuropower: Art in the Age of Cognitive Capitalism', A. De Boever and W. Neidich (ed), *The Psychopathologies of Cognitive Capitalism: Part One*, Berlin: Archive Books.
———. (2014), 'The Mind's Eye in the Age of Cognitive Capitalism', C. Wolfe, T. (ed), *Brain Theory: Essays in Critical Neurophilosophy*, New York, NY: Palgrave Macmillan.
Nitsche, M. (2008), *Video Game Spaces: Image, Play, and Structure in 3D Game Worlds*, Cambridge, MA: MIT Press.
Nusselder, A. (2009), *Interface Fantasy: A Lacanian Cyborg Ontology*, Massachusetts: MIT Press.
Paasi, A. (2011), 'Geography, Space and the Re-emergence of Topological Thinking', *Dialogues in Human Geography*, 1: 299–303.
Papercuts11. (2011), 'Uncharted 2 vs. Uncharted 3 Aiming Tests', http://www.Youtube.com
Parikka, J. (2010), *Insect Media: An Archeology of Animals and Technology*, Minneapolis, MN: University of Minnesota Press.
Parish, J. (2010), 'Final Fantasy's Hiroyuki Ito and the Science of Battle'. Retrieved 31 January 2014, from http://www.1up.com/features/final-fantasy-hiroyuki-ito-science
Patel, A., T. Schieble, M. Davidson, M. C. Tran, C. Schoenberg, E. Delphin and H. Bennett (2006), 'Distraction with a Hand-held Video Game Reduces Pediatric Preoperative Anxiety', *Pediatric Anesthesia*, 16: 1019–1027.
Paul, C. A. (2010), 'Welfare Epics? The Rhetoric of Rewards in World of Warcraft', *Games and Culture*, 5: 158–176.
Pegoraro, R. (2012), 'Gamification: Green Tech Makes Energy Use a Game – and We All Win'. Retrieved 7 April 2014, 2014, from http://arstechnica.com/features/2012/02/gamification-green-tech-makes-energy-use-a-gameand-we-all-win/2/
Persson, M. (2011), 'Terrain Generation, Part 1'. Retrieved 4 April 2014, 2014, from http://notch.tumblr.com/post/3746989361/terrain-generation-part-1
Peters, M. A., E. Bulut and A. Negri (2011), *Cognitive Capitalism, Education, and Digital Labor*, New York, NY: Peter Lang.
PlaystationBlog. (2010), 'Final Fantasy XIII Interview with Battle Director Yuji Abe'. Retrieved 1 July 2014, from http://www.viddler.com/v/ddee7eb1
Pockett, S. (2003), 'How Long Is "Now"? Phenomenology and the Specious Present', *Phenomenology and the Cognitive Sciences*, 2: 55–68.
Poole, S. (2010), 'Edge 93'. Retrieved 9 February 2012, from http://stevenpoole.net/trigger-happy/edge-93/

Postigo, H. (2003), 'From Pong to Planet Quake: Post Industrial Transitions from Leisure to Work', *Information, Communication and Society* 6: 593–607.

Power, M. (2007), 'Digitized Virtuosity: Video War Games and Post-9/11 Cyber-Deterrence', *Security Dialogue*, 38: 271–288.

Prout, A. (1996), 'Actor-network Theory, Technology and Medical Sociology: An Illustrative Analysis of the Metered Dose Inhaler', *Sociology of Health and Illness*, 18: 198–219.

Purchese, R. (2011), 'Battlefield 3 has dedicated servers'. Retrieved 23 September 2014, from http://www.eurogamer.net/articles/2011-02-10-battlefield-3-has-dedicated-servers

Raessens, J. (2014), 'The Ludification of Culture', M. Fuchs, S. Fizek, P. Ruffino and N. Schrape (eds), *Rethinking Gamification*, Leuphana: Meson Press.

Read, J. (2009), 'A Genealogy of Homo-economicus: Neoliberalism and the Production of Subjectivity', *Foucault Studies*, 6: 25–36.

Reichenbach, H. (2012), *The Philosophy of Space and Time*, United States: Dover Publications.

Richardson, I. (2009), 'Sticky Games and Hybrid Worlds: A Phenomenology of Mobile Phones, Mobile Games and the iPhone', L. Hjorth and D. Chan (eds), *Gaming Cultures and Place in Asia-Pacific*, Routledge: part of the Taylor & Francis Group: 213–232.

Richmond, J. (2011), 'A Look at Gun Mechanics in Single-player and Multiplayer'. Retrieved 15 February 2011, from http://www.naughtydog.com/site/post/singleplayer_multiplayer_gun_mechanics/

Roberts, T. (2012), 'From "New Materialism" to "Machinic Assemblage": Agency and Affect in IKEA', *Environment and Planning A*, 44: 2512–2529.

Rogers, T. (2010), 'Final Fantasy XIII'. Retrieved 23 September 2014, from http://www.actionbutton.net/?p=630

Romanillos, J. L. (2008), ' "Outside, It Is Snowing": Experience and Finitude in the Nonrepresentational Landscapes of Alain Robbe-Grillet', *Environment and Planning D: Society and Space*, 26: 795–822.

———. (2013), 'Nihilism and Modernity: Louis-Ferdinand Céline's Journey to the End of the Night', *Transactions of the Institute of British Geographers*, doi: 10.1111/tran.12046.

Rose, G. (2001), *Visual Methodologies: An Introduction to the Interpretation of Visual Materials*, London: Sage Publications.

———. (2003), 'On the Need to Ask How, Exactly, Is Geography "Visual"?', *Antipode*, 35: 212–221.

Rose, G., M. Degen and C. Melhuish (2014), 'Networks, Interfaces, and Computer-generated Images: Learning from Digital Visualisations of Urban Redevelopment Projects', *Environment and Planning D: Society and Space*, 32: 386–403.

Rose, G. and D. Tolia-Kelly, eds (2012), *Visuality/Materiality: Images, Objects and Practices*, Farnham: Ashgate.

Rushkoff, D. (2013), *Present Shock: When Everything Happens Now*, New York, NY: Penguin Group US.

Russell, J. (1993), *Francis Bacon*, London: Thames & Hudson, Limited.

Schull, N. D. (2012), *Addiction by Design: Machine Gambling in Las Vegas*, Princeton, NJ: Princeton University Press.

Shaviro, S. (2010), *Post Cinematic Affect*, Winchester: Zero Books.
Sherry, J. L. (2004), 'Media Effects Theory and the Nature/Nurture Debate: A Historical Overview and Directions for Future Research', *Media Psychology*, 6: 83–109.
Silhavy, T. J., D. Kahne and S. Walker (2010), 'The Bacterial Cell Envelope', *Cold Spring Harbor Perspectives in Biology*, 2: 1–16.
Simondon, G. (1970), 'Entretien sur la mecanologie'. Retrieved 28 July 2013, from http://www.youtube.com/watch?v¼eXDtG74hCL4
———. (1995), *L'individu et sa genese physico-biologique*, Grenoble: Jerome Millon.
———. (2009), 'Technical Mentality', *Parrhesia*, 7: 17–27.
Simonsen, K. (2013), 'In Quest of a New Humanism: Embodiment, Experience and Phenomenology as Critical Geography', *Progress in Human Geography*, 37: 10–26.
Simpson, P. (2009), '"Falling on Deaf Ears": A Postphenomenology of Sonorous Presence', *Environment and Planning A*, 41: 2556–2575.
Slater, M. D. (2007), 'Reinforcing Spirals: The Mutual Influence of Media Selectivity and Media Effects and Their Impact on Individual Behavior and Social Identity', *Communication Theory*, 17: 281–303.
Sloterdijk, P. (2011), *Bubbles: Microspherology*, Los Angeles, CA: Semiotexte/Smart Art.
Smith, M. R. and L. Marx (1994), *Does Technology Drive History?: The Dilemma of Technological Determinism*, Cambridge, MA: MIT Press.
Stiegler, B. (1993), 'Questioning Technology and Time', *Tekhnema*, 1: 31–44.
———. (1998), *Technics and Time, 1: The Fault of Epimetheus*, Stanford, CA: Stanford University Press.
———. (2006), 'Within the Limits of Capitalism, Economizing Means Taking Care'. Retrieved 23 September 2014, from http://arsindustrialis.org/node/2922
———. (2007), 'Anamnesis and Hypomnesis'. Retrieved 12 June 2014, from http://www.arsindustrialis.org/anamnesis-and-hypomnesis
———. (2009a), *Technics and Time, 2: Disorientation*, Stanford, CA: Stanford University Press.
———. (2009b), 'Teleologics of the Snail: The Errant Self Wired to a WiMax Network', *Theory, Culture and Society*, 26: 33–45.
———. (2010a), *Taking Care of Youth and the Generations*, Stanford, CA: Stanford University Press.
———. (2010b), *For a New Critique of Political Economy*, Cambridge: Polity.
———. (2011a), *The Decadence of Industrial Democracies*, Cambridge: Polity.
———. (2011b), *Technics and Time, 3: Cinematic Time and the Question of Malaise*, Cambridge: Polity.
———. (2013a), *Uncontrollable Societies of Disaffected Individuals*, Cambridge: Polity.
———. (2013b), *What Makes Life Worth Living: On Pharmacology*, Cambridge: Polity.
Stiegler, B., C. Venn, R. Boyne, J. Phillips and R. Bishop (2007), 'Technics, Media, Teleology: Interview with Bernard Stiegler', *Theory, Culture and Society*, 24: 334–341.

szefu18. (2011), 'Uncharted 3- Drake's Deception AIM PROBLEM/CONTROLLER LAG/Input Lag – PATCH NEEDED!'. Retrieved 1 July 2014, from https://www.youtube.com/watch?v=K5-9oV9cXwY

Taylor, T. L. (2009), 'The Assemblage of Play', *Games and Culture*, 4: 331–339.

Terranova, T. (2000), 'Free Labor: Producing Culture for the Digital Economy', *Social Text*, 18: 33–57.

———. (2007), 'Futurepublic: On Information Warfare, Bio-racism and Hegemony as Noopolitics', *Theory, Culture and Society*, 24: 125–145.

Thompson, E. (2005), 'Sensorimotor Subjectivity and the Enactive Approach to Experience', *Phenomenology and the Cognitive Sciences*, 4: 407–427.

Thrift, N. (2004), 'Intensities of Feeling: Towards a Spatial Politics of Affect', *Geografiska Annaler: Series B, Human Geography*, 86: 57–78.

———. (2005), 'From Born to Made: Technology, Biology and Space', *Transactions of the Institute of British Geographers*, 30: 463–476.

———. (2008), *Non-Representational Theory: Space, Politics, Affect*, New York, NY: Routledge.

———. (2011), 'Foreword', Y. M. Boutang (ed), *Cognitive Capitalism*, London: Polity.

Tobias, J. (2010), 'Designing Wonder: Complexity Made Simple or the Wii-Mote's Galilean Edge', *Television and New Media*, 11: 197–219.

Turner, M. (1998), *The Literary Mind: The Origins of Thought and Language*, Oxford: Oxford University Press.

Unity. (2014), 'What is Unity?'. Retrieved 30 June 2014, from http://unity3d.com/pages/what-is-unity

Valiaho, P. (2012), 'Affectivity, Biopolitics and the Virtual Reality of War', *Theory, Culture and Society*, 29: 63–83.

Valve. (2012a), 'Hitbox'. Retrieved 3 June 2014, from https://developer.valvesoftware.com/wiki/Hitbox

———. (2012b), 'Source Multiplayer Networking 2012'. Retrieved 7 February 2012, from https://developer.valvesoftware.com/wiki/Source_Multiplayer_Networking

Van Camp, N. (2012), 'From Biopower to Psychopower: Bernard Stiegler's Pharmocology of Mnemotechnologies', *Ctheory.net*, 37: no pagination.

Verbeek, P. P. (2008), 'Cyborg Intentionality: Rethinking the Phenomenology of Human–Technology Relations', *Phenomenology and the Cognitive Sciences*, 7: 387–395.

Verhoeff, N. (2012), *Mobile Screens: The Visual Regime of Navigation*, Amsterdam: Amsterdam University Press.

von Hilgers, P. (2012), *War Games: A History of War on Paper*, Cambridge, MA: MIT Press.

Vradenburg, G. (2014), 'Alzheimer's: Google Glass As Brain Prosthetic'. Retrieved 23 September 2014, from http://www.huffingtonpost.com/george-vradenburg/alzheimers-google-glass-as-brain-prosthetic_b_4637488.html

Wark, M. K. (2009), *Gamer Theory*, Cambridge, MA: Harvard University Press.

Watkins, J., L. Hjorth and I. Koskinen (2012), 'Wising up: Revising Mobile Media in an Age of Smartphones', *Continuum*, 26: 665–668.

Wilson, M. (2012), 'Location-based Services, Conspicuous Mobility and the Location-aware Future', *Geoforum*, 43: 1266–1275.

Winnicott, D. (1953), 'Transitional Objects and Transitional Phenomena – A Study of the First Not-Me Possession', *International Journal of Psycho-Analysis*, 34: 89–97.

Wolf, M., ed. (2008), *The Video Game Explosion: A History from PONG to Playstation and Beyond*, Westport, CT: Greenwood Press.

Wolf, M. J. P. (2001), *The Medium of the Video Game*, Texas: University of Texas Press.

Yee, N. (2009), 'Befriending Ogres and Wood-Elves: Relationship Formation and the Social Architecture of Norrath', *Games Studies* 9.

Yoon, A. (2011), 'Uncharted 3 Dev Defends Changed Controls'. Retrieved 15 November 2011, from http://www.shacknews.com/article/70935/uncharted-3-dev-defends-changed-controls

Zagal, J. P., C. Fernández-Vara and M. Mateas (2008), 'Rounds, Levels, and Waves: The Early Evolution of Gameplay Segmentation', *Games and Culture*, 3: 175–198.

Zichermann, G. and C. Cunningham (2011), *Gamification by Design: Implementing Game Mechanics in Web and Mobile Apps*, Sebastopol, CA: O'Reilly Media, Incorporated.

Zichermann, G. and J. Linder (2013), *The Gamification Revolution: How Leaders Leverage Game Mechanics to Crush the Competition*, New York, NY: McGraw-Hill Education.

ZombiDeadZombi. (2011), 'The Aiming/SP Lag Issue'. Retrieved 1 April 2014, from http://www.gamefaqs.com/boards/615426-uncharted-3-drakes-deception/60878164

# Index

affect/ affective
    auto 42–3
    definition of 23–4
    human/bodily 10, 25–6, 38, 55, 88, 94, 109, 127, 139
    labour 105–6, 108, 147
    negative 6, 46, 107, 108, 117
    positive 6, 12, 44, 107, 115, 117, 125–6
    proletarianization 75
    suspension 107, 111, 112, 115, 117, 118
    technical/ object 42, 76, 143, 146
    technique 104
    transduction 113
    value 7, 13, 105–6
affordance 34
analogue 21–3
anticipation 6, 46, 60, 89, 102 *see also* creative anticipation; protention
artificial intelligence (AI) 19, 29, 57, 96
attention
    control 12, 32, 38, 53, 87, 100, 127
    economy 4–7, 38, 96–7, 102–3, 129–30, 140, 146–7
    envelope power 90–4
    neuropower 38, 72
    psychopower 58–9, 64, 66, 75–8
augmented reality 122–4, 126, 138, 142

*Battlefield III* 12, 40–2, 48
*Beyond: Two Souls* 37
biopower 65–6
Bogost, Ian 26, 29, 122

*Call of Duty 4: Modern Warfare* 11, 73–7
*Call of Duty: Elite* 112
*Call of Duty: Modern Warfare 2* 57
care 5, 109–11, 113–18, 137, 147

*Civilization* 57
code 2, 17–18, 22, 25, 27, 77, 136, 140
cognitive capitalism 6, 38, 40, 53, 105, 124, 129, 140, 146, 147
contingency 6, 10, 55, 94, 103, 115, 131, 140
control society 65–6
creative anticipation 62, 77, 94, 95
culture industries 5, 65, 91–2

Deleuze, Gilles 23, 24, 43, 65–6
*Demon's Soul* 108, 115–17
Digital
    as category of being 21–3, 31
    as computational 1–7, 17–18, 64
    inscription 10
    interface 15–19, 35, 45, 103–5, 122–3, 128–9, 131, 138–40, 142–5
    object 37
    *see also* interface

ecotechnics 13, 105, 108–11, 113, 115, 118, 128
embodiment 11, 19, 24
envelope
    definition of 3, 83–4
    as emergent 9
    envelopment 82–3, 87–8, 94
    as homeomorphic 83, 99
    interface design 32
    Latour, Bruno 119–20
    logics of 54–5, 137–40, 148
    non-videogame 122, 124–30
    videogame 4, 16, 80–4, 88–104, 106, 142–4
    *see also* envelope power
envelope power
    as concept 3–4, 139–40
    as contingent 10

# INDEX

critique 133, 139–40
definition of 90–6
as homeomorphic 83–4
neuropower 6–7, 79, 146–7
non-videogames 118, 119, 122, 124, 129–31, 142
politics 11
psychopower 6–7, 12, 79, 146–7
and sensory capacities 13, 32
videogames 99, 102–4, 108
*see also* envelope

*Final Fantasy* 87, 96
*Final Fantasy IV* 96, 97
*Final Fantasy XII* 97
*Final Fantasy, XIII* 84–90
flat ontology 25–6, 146
Foucault, Michael 65–6
futurity 131

gambling 93–5
game studies 9–11, 140
gamification 13, 119–22, 148
Global Positioning Systems (GPS) 93, 123–4
*God of War* 58
*Grand Theft Auto V* 133

habit
   automatic 72, 89–90
   bodily / embodied 6–7, 19, 37–40, 50, 95, 107, 126
   brain 11–12, 83, 124, 142, 146
   as capacity 97, 103
   as contraction 94
   definition of 53
   and neuropower 37–40, 46, 48
   as practice 83–4
   as skill 134, 139
   and technique 53–5
*Halo 3* 78, 106, 116, 134
*Halo Waypoint* 13, 112
Harman, Graham 7, 8
Heidegger, Martin 2, 8, *see also* care; technicity; temporal (consciousness); temporality (ecstasis); time
homeostasis 36, 42, 98

homogenity
   cultural 92
   temporal 67–73, 126
hypomnesis 65, 132

ideology 9, 20, 21
individuation 10, 39, 91–2
instrumentality 2, 17, 133
interface envelope, definition of 81–4
interface
   as affect 10, 23–4
   as computational device 1–4, 14, 17–18, 78
   as cultural logic and ideology 20–1
   design 64, 70, 98, 127
   digital 15, 21–3, 35, 131, 138, 143–5
   environment 9, 29–31, 34–7, 40–4, 46–7, 57–9, 72, 75–6, 91, 95, 98, 118, 134–5, 139, 142
   games 99, 100–5
   as habit 53–5, 128
   industry 146–7
   map 100, 124
   as object 25–9, 49–50, 140
   as practice 18–19
   radar 101
   smart meter 13, 122, 125, 128–9, 132–3, 136, 144
   smart phone 73, 104, 121, 122–4, 125
   system 13, 84, 86, 108
   theory 16–17
   *see also* digital; envelope

Latour, Bruno 8, 63, 81, 119
*Left 4 Dead* 47

machine zone 93–5
Malabou, Catherine 8, 39, 42, 53
memory
   as automatic 72
   exteriorization of 113, 131–3
   and future 38–9
   human 13, 75, 88–9, 91, 99–100, 102–3, 120, 139–40, 141
   involuntary 43
   neuropower and 93

# INDEX

psychopower and 6, 64, 66
technical / non-human 106
working 38–9
*see also* neuropower;
psychopower; retention;
*Metal Gear* 99
*Metal Gear 2: Solid Snake* 37, 100, 101
*Metal Gear Solid* 99
*Metal Gear Solid 2: Sons of Liberty* 99
*Metal Gear Solid 3: Snake Eater* 99, 101
*Metal Gear Solid 4: Guns of the Patriots* 54, 101
*Metal Gear Solid* V: *Ground Zeroes* 102–3
*Minecraft* 135–7
modulation
    affective 27, 105
    definition of 84
    habitual 54–5
    perceptual 83, 84, 90, 92
    of resolution 37–8, 131
    of space-time 6
    of technicity 89, 131

Nancy, Jean-Luc 8, 15, 34, 108
Neidich, Warren 8, *see also* neuropower
neuroplasticity 39, 53, 92
neuropower
    definition of 6, 37–41
    as distinct from neuropower and envelope power 53–5, 78–9, 90–3, 103, 124, 140, 146
    as technique of control in video games 43–4, 46–50, 72, 75
    *see also* envelope power; habit; Neidich, Warren; psychopower
new materialism 7–8, 144
nonrepresentational theory 11

object orientated ontology 7, 8, 27, 144
objects
    as affective 23–5, 106, 111, 125
    as autonomous units 9
    and consumption 40
    hardware as 64
    as inanimate 7–8
    as inorganically organised 12, 16, 26–7, 107, 139, 143–6

    in interfaces 30–5, 42, 57–8, 84, 121, 123–5, 133, 136
    as non-relational 145
    as relational 142, 145
    resolution of 36–7, 42–9, 50–5, 95, 110, 124, 131, 140
    software 77
    as spheres 82
    as substances with properties 2, 22
    technical 1–3, 5, 10, 13, 19, 28–9, 66, 75, 109, 119
    technicity of 59–63, 95, 108, 110, 124, 131, 140
    as topological 83
    transitional 137
    in videogames 30–5, 42, 57–8, 67, 72–3, 78–9, 86–91, 96, 98–101, 102–3, 135, 141

perpetual now 5, 6, 93, 103, 139
pharmacology 137–8
pharmakon 137, 146
phenomenology 8
*Pong* 106
post-phenomenology 7–8, 143
present-at-hand 61–3, 71, 110, 143
*Prop Hunt 2* 133, 134
protention 64–6, 83–4, 86, 88, 89, 91–5, 98–103, 125–7, 130–2, 134–5, *see also* retention
psychopathology 63
psychopower
    definition of 6, 58–9, 129
    as distinct from neuropower and envelope power 64–7, 92–3, 124, 140, 146
    and interfaces 12, 90–3
    as technique of control in video games 72, 75, 77–9
    *see also* envelope power; neuropower; Stiegler, Bernard

qualities
    definition of 27–9
    as emergent through selective encounters 30, 32, 40, 42, 51, 54, 63, 75, 82, 131, 141–4, 146

primary 27–9
secondary 27–9, 87
as transduced to player 31, 34, 36–7, 43, 46, 48, 61, 83, 88–9, 106, 109, 114, 125

ready-to-hand 60–2, 71, 76, 143
*Red Dead Redemption* 57, 133
representation
    as calculation 87
    and cognition 39
    as gap between image and world 123
    and images 18, 142
    politics 11–12
    power 11–12
    symbolic 16, 31–2, 78
    value 106
    videogames as 20–3
    visual 25–6, 132
*Resident Evil* 98
*Resident Evil 4* 98–9
resolution
    definition of 36–7
    Harman's definition of 33–4
    high 37, 41–55
    of interfaces 16, 95–6, 108, 131, 136
    low 37, 41–55
    Nancy's definition of 35
    and neuropower 12, 40, 78
    of objects 32, 82–4, 124–7, 139–46
    and psychopower 78
    of videogame objects 86–90, 98–100, 101–2, 110, 133
retention
    definition of 64–6
    primary 64–6, 88–9, 91–2, 133
    secondary 64–6, 88–9, 91–2, 95, 99, 103, 123, 127, 131, 132–5, 147
    tertiary 64–6, 86, 93, 99–103, 112, 123, 126–7, 131–4, 138
    *see also* protection

*Sim City 4* 58
Simondon, Gilbert 28, 36, 59
Sloterdijk, Peter 8, 15, 81, 82
smart phone applications 73, 104, 121, 132, 134, 135

software
    computer software 1, 3, 16–20, 26, 34–5, 64, 77–8, 140–1, 145
    mechanisms 47, 49–50, 58
    non-game 121, 127
    open source 135
    studies 9–10, 11
    transduction 37
space / spatial
    as affective 108–9
    bodily 125–6
    as boundary 82, 87–90, 98
    civil 5
    cyber 16, 142
    game 87–90, 100, 120
    as geometric unit 61, 70–1, 75, 77, 113, 130–2, 138–43, 146–7
    as image 123
    materiality of 51–2
    modulation of 103
    and neuropower 93–4
    as non-relation 37
    shared 82
    social 19
    sound and 42
    space-time 3, 6–7, 12, 15–16, 82
    *see also* spheres; topology
speculative realism 7–8, 145
speed
    as difference 74, 76, 141
    in game 58, 85–7
    as non-human 49
    and perception 24
    of player input 36, 47, 70, 86–7, 94–5, 101
    speed-runs 132
spheres 15, 81–2, 112
Stiegler, Bernard 4, 8
*Street Fighter II* 67
*Street Fighter IV* 12, 51–2, 58, 67–8, 70, 72, 73, 77, 78, 94, 139
synaesthesia 43–4, 50, 126

*Team Fortress 2* 47, 78
technicity 146
    as durable fixing of the now 12, 58–64, 77–8, 86
    of game objects 70–5, 86, 88–90, 98

# INDEX

modulation of 16, 95, 131, 140–4
of non game objects 108, 110, 124–7, 133, 136
and psychopower 64–7, 146
as temporal appearance 32, 55, 82–4, 140–4
technics 9, 26, 61
technology / technical/ technological 144
and augmented reality 123, 124, 127, 138
as category of being 2–3, 8–9, 25, 130, 143
determinism 9
digital 129, 133, 139
as ecotechnics 109
as equipmental structure 60, 63
as gigantic 87
and new media 16
as prosthesis 59
as psychopower 75–8
wearable 144
*see also* digital; interface; software; technics; technicity; time
temporal / temporality
consciousness 60, 62–4, 66
ecstasis 57–8, 60–1, 63, 70–2, 76, 108, 141, 147
*see also* time

time
as folding 139, 140–3
in game 57–8, 68
as metrical unit 33, 41, 48–9, 53, 70–2, 116, 118, 126, 132, 146–7
and microdifferentiation 93–7
as process of appearance 55, 108, 140–3
real 102, 128–30
social 5–7
spatialization of 12–13, 67, 71–8, 112–14
*Tomb Raider* 133
topology 83, 88
transduction 16, 28–31, 35, 83, 88, 109, 110–11, 113, 142–3
*Trouble in Terrorist Town* 134

*Uncharted 2* 12, 44
*Uncharted 3* 44–6

virtual 11, 22, 38, 120, 142

*World of Warcraft* 78, 131–2